FE Mechanical Exam Prep

The Complete Guide to Get Ready in No Time - Practice Problems with Detailed Explanations to Conquer the Exam on Your First Attempt with a 98% Pass Rate

Claud Spencer

Copyright © 2024 by Claud Spencer

This document aims to provide accurate and dependable information regarding the covered topic and issue. It is sold under the understanding that the publisher is not obligated to offer accounting, legally authorized, or otherwise qualified services. If professional advice is needed, it is recommended to consult a qualified individual in the respective field. The contents of this document are based on a Declaration of Principles endorsed by both the American Bar Association Committee and a Committee of Publishers and Associations. Reproduction, duplication, or transmission of any part of this document in electronic or printed form is strictly prohibited. Recording or storing this publication is prohibited without written permission from the publisher. The publisher holds no legal responsibility for any errors or misinterpretations resulting from the use or misuse of the policies, processes, or directions provided. The information provided is deemed to be truthful and consistent, with any liability resting solely with the recipient reader. The publisher disclaims any legal responsibility for damages or financial losses arising directly or indirectly from the information contained herein. All copyrights not owned by the publisher belong to their respective authors. The information presented in this document is for informational purposes only and is not binding. The use of trademarks does not imply endorsement or approval by the trademark owner, and all trademarks and brands mentioned are for clarification purposes only, owned by their respective owners and not affiliated with this document.

Table of Contents

Preface 9
   Introduction to the Manual's Structure and Philosophy 9
      Practical Problems and Solutions 9
      Connecting Theory with Real-World Application 9
   How to Maximize Your Use of This Manual 10
      Utilizing the Tools Provided 10

Chapter 1: Navigating the FE Mechanical Exam Landscape 12
   Overview of the Exam 12
      What to Expect on the Exam 12
   Comprehensive Coverage Checklist: Ensuring Alignment with NCEES Standards 13
      Categories and Subtopics: 14
   How to Utilize the NCEES Reference Handbook 17

Chapter 2: Software Applications 19
   Introduction to Mechanical Engineering Software 19
      Overview of Widely Used Software 19
      Fundamentals of AutoCAD 19
      SolidWorks: Modeling and Simulation 21
      ANSYS for Advanced Analysis and Simulation 23

Chapter 3: Mathematical Foundations for Mechanical Engineers 25
   Practical Problems and Solutions in Analytic Geometry and Calculus 27
      Analytic Geometry 27
      Calculus 29
      Practical Application 30
   Mastering Differential Equations Through Application 31
      Theoretical Foundations 31
      Specific Test Strategies 34
   Practice 35

Chapter 4: Statistical Strategies for Engineering Success 37
   Understanding and Applying Key Statistical Measures 38
      Probability Distributions 40

    Regression Analysis 42

  Practice 44

Chapter 5: Ethical Engineering in Practice 46

  Navigating Professional Ethics and Intellectual Property 46

    Intellectual Property: Understanding Your Rights and Responsibilities 46

  Societal Impacts and Responsibilities of Mechanical Engineers 47

  Practice 52

Chapter 6: Economics of Engineering Decisions 54

  Time Value of Money and Cost Analysis in Engineering Projects 54

    Capital Budgeting: Essential Techniques for Engineering Project Valuation 56

    Importance of Risk Assessment in Engineering 56

    Advanced Applications of Cost-Benefit Analysis 58

  Practice 60

Chapter 7: Mastering the Essentials of Electricity and Magnetism 62

  Fundamental Electrical Principles and Their Applications in Mechanical Engineering 62

    Electromagnetism Basics 64

    Electromagnetic Applications in Mechanical Engineering 65

  Detailed Problem-Solving in DC and AC Circuit Analysis 66

    Fundamental Concepts in DC Circuit Analysis 66

    Alternating Current (AC) Circuit Analysis 67

  Practice 69

Chapter 8: Statics and Dynamics: The Mechanics of Movement 71

  Comprehensive Coverage of Statics from Basics to Complex Applications 71

    Types of Forces and Support Reactions 73

    Dynamics of Rigid Bodies 73

    Application in Mechanical Systems 75

  Dynamics, Kinematics, and Vibrations: Solving Real-World Problems. 77

    Vibrations: Oscillatory Motion 78

    Real-World Application: Analyzing a Simple Pendulum 80

    Interactive Section: Real-World Engineering Challenges in Statics and Dynamics 80

  Practice 82

Chapter 9: The Strength of Materials: From Theory to Practical Applications 84

Advanced Topics: Buckling, Indeterminate Systems, and Combined Loading 85

Thermal Stresses in Structural Elements 86
Indeterminate Systems 87
Combined Loading 89

Deeper Understanding of Stress, Strain, and Structural Integrity 89

Stress: Internal Resistance 90
Fatigue and Fracture Mechanics: Understanding Material Behavior Under Cyclic Loading 91
Strain: Deformation of Materials 92
Application Example: Beam Analysis 93

Practice 94

Chapter 10: Advanced Material Properties and Manufacturing Processes 96

In-Depth Look at Material Selection, Properties, and Engineering Processes 96

Fatigue Life Estimation Methods: 99
Mechanical Properties: 101

Corrosion, Failure Mechanisms, and Their Control 102

Controlling Corrosion: Protective Measures and Treatments 102

Practice 105

Chapter 11: Fluid Mechanics for the Practical Engineer 107

Essential Principles and Problem Solving in Fluid Dynamics 107

Fundamental Concepts 107
Dimensionless Numbers and Their Significance in Fluid Dynamics 110
Boundary Layer Theory 112
Hydraulic and Energy Gradient Lines 113
Fluid Properties and Characteristics: 115

Energy, Impulse, and Flow: Tools for Mechanical Engineering 117

Fluid Mechanics for the Practical Engineer 117
Impulse and Fluid Dynamics 118

Practice 120

Chapter 12: Thermodynamics and Heat Transfer: Core of Mechanical Engineering 122

Fundamentals and Applications of Thermodynamics 122

Practical Guide to Heat Transfer 129

    Integrating Heat Transfer Modes 130

  Practice 131

Chapter 13: The Future of Mechanical Engineering- Emerging Technologies and Advanced Materials 134

Revolutionizing Production: Additive Manufacturing 134

    Revolutionizing Industries 135

Automation and Robotics in Manufacturing 136

Chapter 14: Instrumentation, Controls, and Measurements 138

Navigating Complex Control Systems and Dynamic Responses 138

    Stability Analysis 139

Enhancing Precision in Engineering Measurements 139

    Sensors 141

Practice 145

Chapter 15: Design and Analysis for Mechanical Reliability 147

Mechanical Design Principles and Stress Analysis 147

    Reliability Prediction 148

    Environmental Stress Screening (ESS) 149

    Case Studies in Mechanical Design 151

Ensuring Quality and Reliability.. 152

    Reliability Engineering Techniques 152

    Process Capability Analysis:.. 154

Practice 157

Chapter 16: Practical Exam Preparation: Simulated Tests and Solutions.. 159

Full-Length Practice Exams Tailored to the FE Specifications 159

    1EXAM PREP... 159

    2 EXAM PREP.. 174

    3 EXAM PREP.. 189

Solutions 207

    1 EXAM PREP 207

    2 EXAM PREP 211

    3 EXAM PREP 215

- Mastering Test Strategies: Time Management, Identifying Correct Solutions, and Avoiding Pitfalls 219
  - Time Management Techniques 219
  - Identifying Correct Solutions 221
  - Avoiding Common Pitfalls 223
  - Practical Application of Test Strategies 225

Chapter 17: Beyond the FE Exam: Pathways to Professional Engineering 226
- Steps After the Exam: Licensure and Lifelong Learning 226
  - Gaining Professional Licensure 226
  - Embracing Lifelong Learning 226
- Building a Career: Opportunities, Networking, and Continuing Education 227
  - Opportunities in Mechanical Engineering 227
  - Networking for Career Advancement 228
  - Continuing Education 229
  - Building a Career Post-Exam 230

FREE SUPPLEMENTARY RESOURCES

Elevate Your Preparation with Exclusive Free Extra Contents - Utilize the Provided QR Code

- ✔ 500 Practice Problems
- ✔ 15 HRs of E-Learning Content

SCAN THE QR CODE TO DOWNLOAD
COMPLEMENTARY BONUSES

Preface

Introduction to the Manual's Structure and Philosophy

Future engineer, welcome to the right place for achieving success! If you are gearing up for the FE Mechanical Exam, it demonstrates your ambition to exceed what you perceive as your limits. I am both proud and delighted that you have made this pivotal decision and chosen this guide to assist you. While the theoretical aspects are undoubtedly crucial, the most significant challenge often lies in the practical elements of the exam itself. You might wonder, "How can I pass? What should I focus on? How can I streamline my study to avoid unnecessary material?"

This manual answers these questions by optimizing your study time and reducing effort. The reality is, it's not the preparation of theoretical concepts that's daunting but rather mastering the exam itself. This guide focuses on presenting you with all potential challenges you might encounter, complete with detailed, step-by-step solutions to equip you for success.

Practical Problems and Solutions

As I just said in this manual, you will discover a distinctive approach that emphasizes practical problems and solutions, reflecting real-world engineering challenges. Each chapter has been structured to first introduce you to fundamental theories and then immediately apply these concepts through hands-on problem-solving exercises. This methodology ensures that you not only grasp the theoretical underpinnings of mechanical engineering but also develop the competence to apply these principles effectively under exam conditions and in professional practice.

Moreover, the manual is designed to facilitate a deep understanding of core topics mandated by the NCEES, ensuring that every essential subject is covered comprehensively. By integrating complex problem-solving tasks throughout the text, we prepare you to pass the FE Exam and to excel at it.

Connecting Theory with Real-World Application

The manual employs a variety of pedagogical strategies to reinforce learning. For instance, "Theory in Action" segments conclude major sections by illustrating how the discussed principles are employed in real engineering projects, enriching your learning experience and providing a clearer perspective on the material's practical relevance.

This manual represents more than just a study guide; it is a companion in your journey towards becoming a certified professional engineer. It has been crafted with the utmost attention to detail, ensuring that it not only meets but exceeds the rigorous standards of the FE Mechanical Exam. By following this structured approach, you are setting yourself on a path to academic and professional success, equipped with the necessary knowledge and capabilities to thrive in the demanding field of mechanical engineering.

Stay engaged, practice diligently, and utilize this manual as your toolkit to tackle the FE Exam with confidence and precision.

How to Maximize Your Use of This Manual

As your guide through the rigorous journey of the FE Mechanical Exam, it is my privilege to instruct you on how to leverage this manual to its fullest potential. Ensuring complete coverage of NCEES topics and effective exam preparation are not just goals but commitments we make to every reader, particularly driven future engineers like yourself.

Embracing a Strategic Approach to Learning

First and foremost, understand that this manual is structured to align seamlessly with the NCEES exam specifications. Each chapter corresponds directly to a specific section of the NCEES outline, ensuring that no topic is left uncovered. However, mere coverage is not enough; mastery is our ultimate goal. For this reason, as mentioned above, each topic is in dept.

Moreover, I encourage you to actively engage with each section. Do not just read; practice. The end of each chapter includes practice problems that are critical for testing your understanding and preparing you for the types of questions you will face. This active engagement is crucial for converting theoretical knowledge into practical skills.

Utilizing the Tools Provided

Within these pages, you will find numerous tools designed to enhance your learning experience:

- **Detailed Examples**: Theoretical points is illustrated with detailed examples that demonstrate its application in real-world engineering scenarios. These examples are your first step in seeing how abstract concepts manifest in practical tasks.
- **Preparation Strategies:** These strategies are crucial for optimizing your study sessions and are perfectly suited for passing the exam. Here are just a few:
    - Identifying Strengths and Weaknesses: Begin by assessing your knowledge and skills in key areas through diagnostic tests. This allows you to clearly identify which topics you have mastered and which require further study, streamlining your preparation efforts.
    - Planning Your Study: Create a study schedule that prioritizes weaker areas while maintaining proficiency in your strengths. Allocate time blocks for focused study, interspersed with review sessions to reinforce material and track your progress effectively.
    - Memory Techniques: Employ mnemonic devices to enhance retention of complex principles. For example, using acronyms to remember the sequence of operations in a process can simplify recall during exams.
    - Learning Techniques: Engage in active learning by integrating theoretical knowledge with practical application. Techniques like diagramming complex systems or teaching a concept to a peer can solidify your understanding and retention.

How You Can Help This book?

Composing this book has proven to be quite a challenge in fact, debugging for hours feels easier than the process of writing. For the first time in my life, I've encountered writer's block. Understanding the topics is one thing, but attempting to articulate them in a logical, concise, cohesive, and well-organized manner is an entirely different task.

Furthermore, since I've chosen to steer clear of any publishing houses, I can proudly label myself as an "independent author." This is a personal decision that hasn't been without its difficulties, but my dedication to helping others has prevailed.

That's why I would be immensely grateful if you could provide feedback on Amazon. Your input would mean a great deal to me and would go a long way in sharing this material with others. I recommend the following:

1. If you haven't done so already, scan the QR code at the beginning of the book to download the FREE SUPPLEMENTARY RESOURCES.

2. Scan the QR code below and quickly leave feedback on Amazon!

SCAN ME

The optimal approach? Share a short video where you discuss your thoughts on the book! If that feels like too much, there's absolutely no pressure. Providing feedback along with a couple of photos of the book would still be greatly appreciated!

Note: There's no obligation whatsoever, but it would be immensely valued!
I'm thrilled to embark on this journey with you. Are you prepared to delve in?
Enjoy your reading!

Chapter 1: Navigating the FE Mechanical Exam Landscape

Overview of the Exam

Future engineers, the first critical step in your journey starts now. As we delve into the Fundamentals of Engineering (FE) Mechanical Exam, it's essential to understand not only the structure and requirements of the exam but also the strategic approach necessary for success.

The FE Mechanical Exam, is designed to test your knowledge and aptitude in fundamental mechanical engineering concepts. The exam serves as a pivotal benchmark for recent graduates and young professionals wishing to advance into responsible positions in the field of engineering.

Why FE Mechanical Exam Is Important

As You certainly know, the Exam is typically the first step in obtaining professional engineering licensure. By passing this exam, you demonstrate a comprehensive understanding of mechanical engineering fundamentals, which is crucial for your advancement to the next stage of professional practice. Moreover, obtaining this certification is a milestone, a testament to your commitment and capability within the engineering community.

What to Expect on the Exam

The FE Exam is a computer-based test that comprises 110 multiple-choice questions. You are allotted 6 hours to complete the exam, which includes a scheduled break and time for a tutorial on how to use the exam interface. The questions are crafted to gauge your understanding in the following areas:

- **Mathematics and Statistics**
- **Ethics and Professional Practice**
- **Engineering Economics**
- **Electricity and Magnetism**
- **Statics and Dynamics**
- **Mechanics of Materials**
- **Fluid Mechanics and Thermodynamics**
- **Heat Transfer**
- **Material Properties and Processing**
- **Instrumentation, Controls, and Mechanical Design**

The first tips that I'm going to give you is this: it's vital to approach your study with a focus on breadth rather than depth, as the FE Exam covers a wide range of topics at a fundamental level.

How to Prepare Effectively

Effective preparation for the FE Mechanical Exam involves a blend of theoretical study and practical application. This manual is designed to guide you through both aspects:

- **Theoretical Study**: Each chapter provides a thorough exploration of key concepts, ensuring that you grasp the foundational theories essential for the exam.
- **Practical Application**: Accompanying the theoretical discussions, you will find practical problems and scenarios that simulate real-world applications and exam-like questions. This approach helps bridge the gap between knowing the material and applying it effectively under exam conditions.

Comprehensive Coverage Checklist: Ensuring Alignment with NCEES Standards

As we progress through our journey to prepare you for the FE Mechanical Exam, it is essential to align our study strategies closely with the standards defined by the National Council of Examiners for Engineering and Surveying (NCEES). This section of our manual, "Comprehensive Coverage Checklist," is designed to ensure that you, as a future engineer, are thoroughly prepared across all required topics outlined by the NCEES for the mechanical engineering discipline.

Why Comprehensive Coverage Is Crucial

The FE Exam is comprehensive, encompassing a broad spectrum of topics from mathematics and statistics to complex mechanical systems. NCEES has meticulously outlined these topics to ensure that candidates are tested on all fundamental aspects of mechanical engineering necessary for entry-level proficiency. Our checklist serves as a roadmap to guide your studies and ensure that no essential topic is overlooked.

How the Checklist Is Structured

The checklist in this manual is structured to mirror the NCEES FE Mechanical Exam specifications directly. For each major category listed by the NCEES, we provide a detailed breakdown of subtopics covered in this manual.

Implementing the Checklist

Step-by-Step Alignment:

- **Identify Each Topic:** Start by familiarizing yourself with the broad categories and corresponding subtopics as outlined by the NCEES. This manual has a chapter or a section dedicated to each category, ensuring a holistic approach to your study.

- **Check for Completeness:** As you proceed through each chapter, use the checklist to mark topics you have studied. This practice will help you track your progress and identify any areas that may need additional review.

Future Insights

While this chapter lays the groundwork for ensuring comprehensive coverage, more detailed strategies for mastering each topic will be provided in the specific chapters that follow. This includes not only learning the material but also applying it effectively in exam-like scenarios through practice problems and simulations, which are critical for translating knowledge into success on the exam day.

Remember, the goal of this manual is not only to prepare you for the Exam but also to build a solid foundation in mechanical engineering principles that will support your career long after the exam is passed.

Categories and Subtopics:

1. **Mathematics and Statistics**
   - Analytic Geometry
   - Calculus (Differential, Integral, Multivariable)
   - Ordinary Differential Equations
   - Linear Algebra (Matrix operations, Vector analysis)
   - Numerical Methods
   - Probability and Statistics (Distributions, Measures of Central Tendencies)

2. **Ethics and Professional Practice**
   - Codes of Ethics
   - Public Health, Safety, and Welfare
   - Intellectual Property
   - Societal Considerations

3. **Engineering Economics**
   - Time Value of Money
   - Cost Analysis
   - Economic Decision Making

4. **Electricity and Magnetism**
   - Electrical Fundamentals
   - DC Circuit Analysis

- AC Circuit Analysis
- Motors and Generators

5. **Statics**
   - Force Systems
   - Equilibrium of Rigid Bodies
   - Frames and Trusses
   - Centroids and Moments of Inertia
   - Static Friction

6. **Dynamics, Kinematics, and Vibrations**
   - Kinematics of Particles
   - Kinetics of Rigid Bodies
   - Work-Energy and Impulse-Momentum Theorems
   - Vibrations

7. **Mechanics of Materials**
   - Stress and Strain Analysis
   - Shear and Moment Diagrams
   - Combined Loading
   - Buckling of Columns

8. **Fluid Mechanics**
   - Fluid Properties
   - Fluid Statics and Dynamics
   - Energy, Impulse, and Momentum Considerations

9. **Thermodynamics and HVAC**
   - Energy Transfers
   - Thermodynamic Laws
   - Refrigeration Cycles
   - HVAC Systems

10. **Material Properties and Processing**
    - Material Properties (Mechanical, Thermal)
    - Manufacturing Processes
    - Phase Diagrams and Heat Treatment

11. **Heat Transfer**
    - Conduction, Convection, Radiation
    - Heat Exchangers

12. **Instrumentation and Controls**
    - Measurement Techniques
    - Control Systems

13. **Mechanical Design and Analysis**
    - Stress Analysis
    - Failure Theories
    - Machine Component Design

Understanding the NCEES Reference Handbook

The NCEES Reference Handbook is the only informational resources allowed during the exam. It contains all the equations, charts, and tables you'll need to reference to solve the exam questions. Familiarity with this handbook is not just beneficial; it's essential. Knowing where to find necessary information quickly can save valuable time and reduce stress during the exam.

Efficient Use of the Handbook

Navigating the Handbook: Start by downloading the latest version of the NCEES Reference Handbook from the official NCEES website. It's crucial to use the version that will be provided during your exam, as updates can include critical changes to the content.

Familiarization with Content:

- **Skim through each section** to get an overview of the contents and organization.
- **Identify key sections** relevant to mechanical engineering—such as thermodynamics, fluid mechanics, and material science—which you'll likely refer to during the exam.

Moreover, integrating your study sessions with the handbook is vital. As you review each topic covered in this manual, refer to the corresponding sections in the handbook. This practice not only aids in retention but also ensures you are comfortable navigating the handbook quickly during the exam.

How to Utilize the NCEES Reference Handbook

Strategic Tips for Handbook Mastery

- **Create a mental map of the handbook's layout.** Know which topics are covered in which sections. This mental mapping will help you quickly locate necessary formulas or data during the exam.
- **Practice with purpose.** When working through practice problems, always use the handbook as your reference source to simulate the exam experience.
- **Use bookmarks and notes** during your preparation phase to mark important pages or sections in the digital PDF. Though bookmarks aren't available during the actual exam, the practice of noting where critical information is located can be beneficial.
- **Real-Time Search Drills**: Regularly practice locating information as quickly as possible. Set up drills where you challenge yourself to find specific information in the handbook within a minute. This practice will sharpen your search skills and reduce the time you spend looking for information during the actual exam.
- **Leverage Technology for Simulation**: Utilize PDF reader tools that allow for advanced searching techniques, including keyword highlighting and indexed search options. Familiarize

yourself with these tools during your study sessions to mimic the digital interface you will encounter during the exam.

During the Exam:

- **Quick navigation is key.** Utilize the handbook's search function effectively—familiarize yourself with the keywords and phrases associated with different topics to speed up your search process.

- **Think Ahead**: Anticipate what information you might need next while solving a problem. This forward-thinking approach can save time and reduce the back-and-forth in the handbook.

Chapter 2: Software Applications

Introduction to Mechanical Engineering Software

Software tools is essential in this field. These tools not only enhance your design capabilities but also integrate seamlessly into every phase of the engineering workflow. Let's explore three of the most widely used software in the field: AutoCAD, SolidWorks, and ANSYS, each serving distinct purposes and bringing unique strengths to your projects.

Overview of Widely Used Software

AutoCAD: Starting with the first one, a robust drafting tool developed by Autodesk. It is renowned for its precision in creating detailed 2D and 3D models. Whether you are drafting a small part or an entire assembly, AutoCAD's versatile toolset allows you to render geometric shapes, manage properties, and organize data with exceptional accuracy. Its application extends from simple sketches to complex architectural plans, making it a staple in both educational and professional settings.

SolidWorks: Moving on to SolidWorks, which is primarily known for its powerful 3D modeling capabilities. Developed by Dassault Systèmes, SolidWorks excels in creating intricate designs and simulating real-world conditions for those designs to test their viability. From drafting initial concepts to executing detailed assembly analysis, this software offers tools for every stage of design. Its user-friendly interface and extensive simulation options make it invaluable for tasks that require advanced engineering analysis such as stress tests and the dynamic behavior of components.

ANSYS: Lastly, ANSYS specializes in comprehensive analysis and simulation. It stands out for its ability to perform finite element analysis (FEA) and computational fluid dynamics (CFD), crucial for predicting how products will perform under various physical conditions. ANSYS helps in identifying potential problems before prototypes are built, saving time and resources. It is particularly useful in fields requiring detailed understanding of environmental impacts on materials, such as aerospace, automotive, and civil engineering.

As you can imagine the impact of these software tools on the engineering discipline cannot be overstated. By leveraging AutoCAD, SolidWorks, and ANSYS, you can achieve a level of precision that manual calculations and older methods simply cannot match. This precision ensures that designs are accurate and reliable, significantly reducing the margin of error Furthermore, the ability to simulate and analyze in a virtual environment accelerates the development process. You can iterate designs faster, perform more comprehensive tests, and refine systems without the high costs and time delays associated with physical prototyping.

These software tools also improve efficiency and productivity by automating routine tasks and allowing more time for innovation and problem-solving. As you advance in your career, the skills to utilize these tools effectively will enable you to lead projects that meet stringent standards, are economically viable, and can be delivered within tighter deadlines..

Fundamentals of AutoCAD

AutoCAD, we reiterate that, is an indispensable tool for creating precise 2D and 3D drawings and models in mechanical engineering. Its comprehensive set of features allows you to draft, visualize, and edit geometric shapes with precision.

AutoCAD's user interface might seem daunting at first, but it is designed to offer you efficient access to its powerful features. The workspace is customizable, allowing you to tailor the toolbars and command settings to your frequent needs. As you start, you'll interact mainly with the command line, toolbars, and the drawing area where the magic happens.

Getting more specific, the software allows you to work by layers organizing different elements of your drawings. Each layer can represent different components of a mechanical assembly or different types of information like dimensions, centerlines, or material specifications. Learning to manage layers effectively ensures that your drawings remain organized and easy to navigate.

Imagine a scenario where you are tasked with designing a gearbox for a high-performance racing car. The gearbox must be compact, efficient, and capable of withstanding high torque and speed. Using AutoCAD, you begin by drafting the initial layout, placing each gear and bearing according to the design specifications. Through iterative modifications and utilizing precise measurement tools, you adjust the positions and dimensions of each component to optimize the gearbox's spatial configuration and ensure smooth operation.

In another instance, consider the design of a custom fixture intended for an automated assembly line. The fixture needs to hold various components securely during the assembly process. Starting with a basic sketch in AutoCAD, you develop the fixture's design by adding clamps and support structures tailored to the dimensions and shapes of the components.

You can see how AutoCAD serves as a critical tool in the toolbox of a mechanical engineer. It not only aids in the design and execution of projects but also sharpens your problem-solving skills, preparing you to tackle diverse challenges in the mechanical engineering domain. By mastering AutoCAD, you equip yourself with the skills to translate theoretical concepts into tangible, functional designs, bridging the gap between idea and reality in the engineering world.

SolidWorks: Modeling and Simulation

Introduction

SolidWorks is a state-of-the-art 3D CAD (Computer-Aided Design) software extensively used in mechanical engineering to build precise models and assemblies. The software's robust functionality includes everything from basic geometric modeling to advanced structural and motion analysis. The core strength in this software lies in its ability to construct detailed, accurate 3D models and assemblies. This, indeed, provides a seamless transition from simple parts to complex machines, facilitating both the design and visualization processes. What sets SolidWorks apart is its integrated simulation suite, which allows you to perform stress tests, dynamic analysis, and fluid dynamic simulations directly on the model. These features are crucial for virtual prototyping, significantly reducing the need for physical prototypes and enabling faster iterations.

Essential for Prototyping and Testing

Let's think about this for a moment, Using SolidWorks, it is possible to simulate real-world conditions to see how a product will behave in the presence of various stresses and motions; without this functionality, we will not be able to ensure that we identify potential problems early in the design process, having to forgo product improvement, reliability, and safety before production begins.

Practical Tutorials in SolidWorks

To truly grasp the capabilities of SolidWorks, engaging with hands-on tutorials is essential. Let's walk through the modeling of a common mechanical component—an industrial gear—and how to perform simulations to analyze its performance under operational stresses.

Modeling a Mechanical Gear:

1. **Starting the Design**: Launch SolidWorks and create a new part. Begin by selecting the front plane to sketch. Use the line and arc tools to draw the cross-sectional profile of a gear tooth. This sketch will be the foundation of your gear's structure.

2. **Creating the Gear Body**: With your sketch complete, activate the Revolve Boss/Base feature. Revolve the sketch around its central axis to form a solid 3D model of the gear. Ensure that the dimensions align with the required specifications for your specific application.

3. **Adding Details**: To enhance the gear, use the Fillet feature. Apply fillets to the edges of the gear teeth to round them off. This step is crucial as it helps reduce stress concentrations, which in turn extends the gear's operational life.

4. **Refining the Gear Teeth**: Adjust the profile of the teeth to ensure they mesh smoothly with mating gears. Use the Sketch Fillet tool to create precise curves where needed, enhancing the durability and performance of the gear.

5. **Adding a Central Hole**: Sketch a circle at the center of the gear on the front plane. Use the Extruded Cut feature to create a hole through the center, allowing the gear to fit onto a shaft.

6. **Implementing Advanced Features**: For more complex gears, consider using features such as the GearMate tool to simulate the interaction with other gears. This allows you to check for proper meshing and alignment in a virtual assembly.

7. **Applying Material Properties**: Assign appropriate material properties to the gear to simulate its real-world behavior. Choose materials based on the gear's operational environment and load conditions to ensure accurate performance predictions.

8. **Performing a Stress Analysis**: Use the Simulation tools in SolidWorks to run a stress analysis on the gear. Identify any areas of high stress and refine the design as needed to improve durability.

9. **Finalizing the Design**: Review the entire gear model, making any necessary adjustments. Ensure all dimensions, features, and material properties are correct. Save the final model and prepare it for manufacturing or further simulation.

10. **Documenting the Design**: Create detailed technical drawings of the gear. Include all necessary views, dimensions, and tolerances. These drawings are essential for manufacturing and quality control.

Conducting Stress Analysis:

Now, let's see how to conduct a stress analysis on the gear to ensure its durability and performance under load conditions.

1. **Setting Up the Simulation**: Begin by navigating to the Simulation tab in SolidWorks. Initiate a new study by selecting the type of simulation you require—either static or dynamic. For our purposes, we'll focus on a static stress analysis to assess the gear's strength under maximum load conditions. This type of analysis is generally sufficient for evaluating how the gear will perform under steady operational loads.

2. **Defining Material Properties**: Before applying loads, ensure that the material properties of the gear are correctly defined. This involves selecting the appropriate material from the SolidWorks material library, which includes properties like Young's modulus, Poisson's ratio, and yield strength. Accurate material properties are crucial for reliable simulation results.

3. **Applying Loads and Fixtures**: Now, apply the necessary forces and constraints to the model. First, simulate the torque the gear will experience by applying a rotational force. This can be done by selecting the appropriate faces or edges of the gear. Next, fix the gear's axis to replicate its actual mounted position in the machinery, ensuring it reflects real-world conditions.

4. **Setting Contact Conditions**: If the gear interacts with other components, define the contact conditions. This includes specifying whether the contact is bonded, no penetration, or another relevant type. Properly defining these conditions ensures that the simulation accurately reflects the gear's operational environment.

5. **Meshing the Model**: Generate a mesh for the gear model. A finer mesh can increase the accuracy of the results, especially around areas of high stress. However, it's important to balance detail with computational efficiency. SolidWorks provides automatic meshing tools, but manual adjustments might be necessary for complex geometries.

6. **Running the Simulation**: With all parameters set, execute the stress analysis. SolidWorks will calculate the stress distribution across the gear, taking into account the applied loads and constraints. This process may take some time depending on the complexity of the model and the fineness of the mesh.

7. **Reviewing Results**: Once the simulation is complete, review the results displayed by SolidWorks. Examine the stress distribution, particularly focusing on areas highlighted in red, which indicate high stress and potential failure points. Use the results to identify any design weaknesses.

8. **Interpreting the Data**: Analyze the data to understand how the gear will behave under operational loads. Look at the maximum stress values and compare them with the material's yield

strength. If the stress exceeds the material's limits, consider redesigning the gear to improve its performance and safety.

9. **Optimizing the Design**: Based on the analysis, make necessary adjustments to the gear design. This might involve changing the geometry, altering the material, or adding fillets to reduce stress concentrations. Re-run the simulation to verify the effectiveness of these changes.
10. **Documenting Findings**: Finally, document your findings and design adjustments. Create a detailed report that includes images of the stress distribution, key data points, and any design changes made. This report is essential for communicating your analysis and ensuring the gear's reliability in practical applications.

ANSYS for Advanced Analysis and Simulation

ANSYS is THE company stands at the forefront of engineering simulation software, offering powerful tools for Computational Fluid Dynamics (CFD) and Finite Element Analysis (FEA). These tools are crucial for your growth as they allow for the exploration of physical phenomena without the need for physical prototypes.

Basics

ANSYS is a comprehensive simulation platform used extensively in the engineering community. The software is designed to help predict a product's performance in the real world using a multitude of simulation options that cover a wide range of physics, including structural, thermal, and fluid dynamics.

FEA and CFD Capabilities: ANSYS excels in providing detailed insights through FEA, which helps in determining the impact of forces on various structures. This analysis makes it clear where a project might fail or how it could be improved. Similarly, ANSYS's CFD capabilities allow you to simulate the flow of fluids and heat transfer in and around your designs, which is essential for optimizing and validating HVAC systems, aerodynamics in automotive designs, and more.

Engineering Specific Modules: For mechanical engineers, ANSYS offers specialized modules that enhance your ability to design, simulate, and refine components and systems. These modules enable precise simulations of material behavior under different stress levels, temperatures, and physical forces, providing a holistic view of potential design challenges before they become costly.

Guided Tutorials

To effectively harness the capabilities of ANSYS, it's beneficial to engage in guided tutorials that offer a step-by-step approach to setting up and running simulations. Let's explore how to set up a CFD simulation for a cooling system, a common yet complex task for mechanical engineers.

Setting Up Fluid Dynamics in a Cooling System:

1. **Initialize the Simulation Workspace:** Open ANSYS and select the fluid dynamics module. Create a new project and import your model of the cooling system.
2. **Define Material Properties:** Specify the type of fluid, its temperature, flow characteristics, and any particulates or impurities that might affect fluid flow.
3. **Set Boundary Conditions:** Define the inlet and outlet conditions for the fluid, such as velocity, pressure, and temperature, ensuring they mirror real-world operational settings.

4. **Mesh the Model:** Apply a mesh to the model, which divides the space into discrete elements that can be individually analyzed. A finer mesh can lead to more accurate results but requires more computational power.
5. **Run the Simulation:** Execute the simulation to observe how the fluid moves through the cooling system, identifying areas where there may be heat build-up or insufficient flow.

Understanding the application of ANSYS in real-world projects can significantly enhance your comprehension of its practical benefits.

Consider a scenario where a leading tech company uses ANSYS to perform thermal analysis on a new smartphone. The engineers use ANSYS to simulate how different design changes affect heat dissipation. The analysis helps them to optimize the placement of components to minimize overheating without compromising on performance.

Through this simulation, the team successfully reduces the risk of overheating by 20%, ensuring the product's reliability and enhancing user satisfaction.

Chapter 3: Mathematical Foundations for Mechanical Engineers

	Formula	Description	Graphical Representation
Analytic Geometry	$x^2 + y^2 = r^2$	Equation of a circle with radius r.	
	y=mx+b	Equation of a line with slope m and intercept b.	
Differential Calculus	$f'(x) = \frac{dy}{dx}$	Derivative of a function y=f(x)	
	$\frac{d}{xd}[x^n] = nx^{n-1}$	Power rule for differentiation.	
Integral Calculus	$\int f(x)dx$	Indefinite integral of f(x)	
	$\int_a^b f(x)dx$	Definite integral from a to b.	
	$\int x^n dx = \frac{x^n + 1}{n+1} + C$	power rule for integration.	

Fundamental Theorem of Calculus	$\int_a^b f'(x)dx = f(b) - f(a)$	Relates differentiation and integration.	
Ordinary Differential Equations (ODEs)	$\frac{dy}{dx} + P(x)y = Q(x)$	First-order linear differential equation.	
Partial Differential Equations (PDEs)	$\frac{\partial u}{\partial t} = c^2 \frac{\partial^2 u}{\partial x^2}$	Heat equation (one-dimensional). Transforming a function to s-domain.	
Laplace Transforms	$L\{f(t)\} = \int_0^\infty e^{-st} f(t)dt$	Transforming a function to s-domain	
Fourier Transforms	$F\{f(t)\} = \int_{-\infty}^\infty f(t)e^{-i\omega t}dt$	Decomposing function into sinusoids.	
Multivariable Calculus	$\nabla f = \left(\frac{\partial f}{\partial x}, \frac{\partial f}{\partial y}, \frac{\partial f}{\partial z}\right)$	Gradient of a scalar field f.	
Vector Calculus	$\nabla \cdot F = \frac{\partial F1}{\partial x} + \frac{\partial F2}{\partial y} + \frac{\partial F3}{\partial z}$	Divergence of a vector field F.	
		Curl of a vector field F	

Practical Problems and Solutions in Analytic Geometry and Calculus

Theoretical Foundations

Analytic Geometry and Calculus form the core tools for modeling and solving real-world engineering problems. Understanding these subjects deeply is essential, not only for the FE Exam but for your future engineering challenges.

Analytic Geometry

This branch of mathematics provides a bridge between algebraic equations and geometric curves or surfaces. For mechanical engineers, proficiency in this area enables the visualization and analysis of objects in two and three dimensions, which is crucial for design

and analysis tasks.

Historical Context and Evolution: Analytic Geometry and Calculus have deep historical roots that significantly influenced modern engineering. Developed by René Descartes and Isaac Newton, among others, during the 17th century, these fields provided the mathematical framework necessary for the precise analysis and representation of physical spaces. Analytic geometry, merging algebra with geometric interpretation, allowed for the visualization of complex shapes through equations, fundamentally transforming architectural and engineering design methods. Calculus introduced concepts of limits, derivatives, and integrals, empowering engineers to model dynamic systems and changes, such as motion dynamics and material stresses, which are critical for designing everything from bridges to engines. These mathematical tools have been pivotal in advancements across all engineering disciplines, underpinning the development of technologies from the Industrial Revolution to contemporary computational engineering.

Fundamentals This field utilizes the Cartesian coordinate system to explore the relationships between algebraic equations and geometric figures. By representing geometric shapes such as lines, circles, and curves through equations, engineers can precisely calculate distances, angles, intersections, and other relevant properties. For example, the formula for a circle in the plane, $x^2 + y^2 = r^2$, not only defines its shape but also allows engineers to solve for intersections or tangent lines analytically.

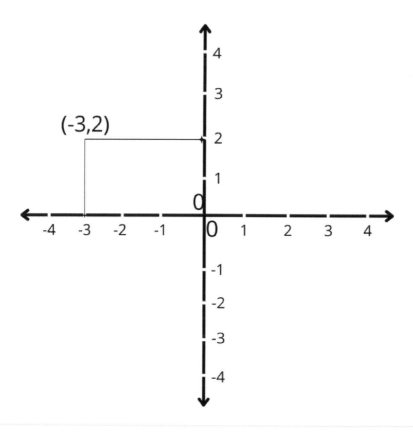

Application in Mechanical Engineering: In mechanical engineering, Analytic Geometry is essential for designing and visualizing complex components and assemblies. It is used extensively to shape the contours of automotive parts and to map out stress distributions in structures under load, leveraging both geometric and algebraic principles to enhance accuracy and effectiveness in engineering tasks.

Complex 3D Modeling: Beyond two dimensions, Analytic Geometry extends to three dimensions, where it plays a pivotal role in the field of computer-aided design (CAD) and other simulation software. In 3D space, the ability to calculate the equations of planes and volumes, and to understand their interactions, is critical for constructing models that can be tested under real-world conditions. For instance, determining the intersection of a plane with a cylinder might be necessary to ascertain the fit and function of a mechanical joint.

Analytical Problem Solving: Furthermore, Analytic Geometry equips engineers with the tools to tackle optimization problems, such as finding the shortest distance between points or the optimal shape for a component to withstand force. These calculations often involve derivatives and integrals, connecting directly with calculus to provide a comprehensive toolkit for quantitative analysis in engineering tasks.

Calculus

Calculus, a fundamental branch of mathematics, is instrumental in mechanical engineering for analyzing and modeling dynamic systems. It is divided into two main branches: differential calculus and integral calculus, each addressing different but complementary aspects of mathematical analys

Differential Calculus: Understanding Rates of Change Differential calculus centers on the concept of the derivative, which quantifies the rate at which a function varies with respect to its input. In mechanical engineering, this translates to analyzing how physical quantities such as velocity, force, and energy change over time or space.

The velocity, derived by differentiating the position function relative to time, indicates the rate at which an object's position is changing as time progresses. The derivative also aids in optimizing engineering processes. Engineers can identify points where a function's derivative (rate of change) is zero to find where a system's performance may be maximized or minimized. This is crucial for optimizing tasks like minimizing material use while ensuring maximum structural integrity.

$$v(t) = \frac{dx}{dt}.$$

Integral Calculus: Measuring Accumulation On the other hand, integral calculus deals with the concept of integration, which can be thought of as the inverse process of differentiation. It is concerned with accumulation, such as calculating the area under a curve, the total displacement traveled given a velocity function, or the total energy expended over a time interval. $\int_a^b f(x)\, dx$

Mechanical engineers, use it to determining quantities like the total heat transfer in a system, which involves integrating the heat transfer rate over time. It also applies to material science, for example, in calculating the mass of an object when its density varies across its volume. Integrals help in computing work done by a variable force and the center of mass of complex shapes, which are integral to engineering analyses and applications.

Connecting Differential and Integral Calculus The Fundamental Theorem of Calculus links these two branches by stating that differentiation and integration are inverse processes. In practical terms, this means that if you integrate a rate of change (derivative), you can find the total change over an interval, and vice versa. This relationship is pivotal in solving differential equations that model physical phenomena in mechanical engineering, such as thermal dynamics, fluid flow, and material deformation.

Advanced Problem-Solving Techniques: Advanced problem-solving in mechanical engineering often employs multivariable and vector calculus to handle complex systems.

- Multivariable calculus allows engineers to analyze changes across multiple dimensions simultaneously, crucial for understanding fluid dynamics and thermal properties in the actual scenario.

- Vector calculus, besides, is essential for dealing with vector fields and forces, enabling engineers to calculate field strengths and directions crucial for electromagnetic and aerodynamic applications.

These techniques support the creation of more sophisticated models and simulations, enhancing the design and functionality of engineering projects. Such tools are indispensable for optimizing design and troubleshooting potential issues in advanced mechanical systems.

Practical Application

To truly master these concepts, you must engage with them practically. Therefore, this section includes a variety of problems that reflect typical scenarios you might encounter both in the exam and in professional practice.

Problem Set Overview:

- **Curve Sketching:** Utilize derivatives to determine the critical points and behavior of functions, a skill vital for understanding material stress-strain curves.
- **Optimization Problems:** Solve problems that require finding maximum or minimum values, common in design and cost-efficiency studies.
- **Integration Techniques:** Apply integral calculus to calculate areas under curves, volumes of solids of revolution, and other physical quantities.

Specific Test Strategies

As you work through these problems, keep the following strategies in mind:

- **Understand the Problem:** Carefully read each problem to fully understand what is required before beginning calculations.
- **Draw Diagrams:** Whenever possible, sketch diagrams to visualize the problem. This can often provide insights into how to approach the solution.
- **Check Units:** Always ensure that your final answer has the correct units. This is a simple but crucial check that can prevent errors.

This structured approach, emphasizing both theory and practice, is designed to build a robust foundation in mathematical concepts crucial for mechanical engineers. As we progress through this manual, these concepts will be further applied to more complex engineering problems, ensuring you are well-equipped for both the Exam and professional practice.

Mastering Differential Equations Through Application

Differential equations are at the heart of engineering as they describe how physical quantities change over time and space, providing a foundation for modeling dynamic systems like mechanical vibrations, heat transfer, and fluid flow.

Theoretical Foundations

Differential equations can be broadly categorized into ordinary differential equations and partial differential equations, each serving distinct purposes in mechanical engineering:

- **Ordinary Differential Equations (ODEs)** deal with functions of one variable and their derivatives. They are essential for modeling systems where changes occur concerning a single independent variable, such as time.

$$m\frac{d^2x}{dt^2} + c\frac{dx}{dt} + kx$$

- **Partial Differential Equations (PDEs)** include several independent variables and their partial derivatives. They are crucial in scenarios where change occurs across more than one dimension, such as temperature distribution across a surface or fluid flow in a pipe.

$$\frac{\partial u}{\partial t} = \alpha \frac{\partial^2 u}{\partial x^2},$$

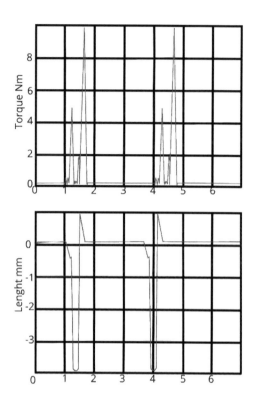

While we have just outlined the basic distinctions between ordinary differential equations and partial one, higher-order differential equations, which involve derivatives beyond the first degree, are essential for modeling more complex dynamic systems. These equations can describe phenomena where the present state is influenced by multiple past rates of change, offering a more comprehensive view of mechanical systems' behaviors under various forces and conditions. This sophisticated approach is critical for accurately simulating real-world scenarios in mechanical engineering, such as advanced vibrations and multi-body dynamics

Linear and Nonlinear Differential Equations: Linear differential equations, which assume linearity in the relationship between variables, are simpler to solve and often used for systems where interactions do not depend on the magnitude of variables. In contrast, nonlinear differential equations accommodate more complex interactions, where outputs are not proportional to inputs, enabling more accurate modeling of real-world phenomena like turbulence and material fatigue. These equations can reflect the non-linear behavior of systems under large perturbations or in highly dynamic environments, which we will explore through specific examples later in this section.

Let's explore how these equations allow us to describe the behavior of evolving physical systems, providing the necessary tools for a detailed and precise analysis of engineering dynamics.

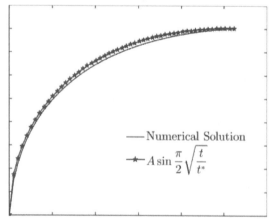

Analytical vs. Numerical Solutions: Analytical solutions provide exact answers using algebraic methods and are suitable for simpler differential equations where such solutions exist. However, in real-world engineering problems, equations often become too complex for analytical solutions. In these cases, numerical methods, such as Runge-Kutta methods or Euler's method, are employed. These techniques approximate solutions and are indispensable for handling nonlinear differential equations or systems where exact solutions are not feasible.

Stability and Control: The stability of a system denotes its capacity to return to equilibrium after experiencing a disruption. Differential equations help predict the conditions under which a system remains stable or becomes unstable. In control systems, they are used to design feedback loops that maintain system stability under changing conditions.

Transform Methods: These mathematical tools are particularly powerful for analyzing systems where vibrations and heat transfer are involved.

Laplace transforms convert differential equations from the time domain, including initial conditions, into simpler algebraic equations in the s-domain. Conversely, Fourier transforms are essential for decomposing complex oscillatory systems into sinusoids, which aids in the analysis of signal processes and heat distribution in materials. For instance, Laplace transforms are instrumental in electrical circuit analysis where they help simplify circuits with switches by converting differential equations into algebraic ones, making the solution process more manageable. Similarly, Fourier transforms are extensively used in mechanical engineering to analyze the vibrational components of a machine. They decompose complex vibrations into simpler sinusoidal waves, enabling engineers to identify and address potential resonant frequencies that could lead to structural failures.

Practical Application

To effectively master these concepts, you must engage directly with practical problems that reflect real-world engineering challenges:

Example Problems:

1. **Modeling a Damped Harmonic Oscillator:** Use a second-order ODE to model the motion of a spring-mass-damper system, focusing on how the damping coefficient affects the system's behavior over time.
2. **Heat Equation:** Apply a PDE to solve a heat transfer problem in a rod, determining the temperature distribution along the rod over time given specific boundary conditions.

Specific Test Strategies

As you delve into these problems, remember to:

- **Dimensional Analysis:** Always check the units of your results. Consistent units throughout your calculations are crucial for correctness.
- **Utilize Computational Tools:** Learning to solve differential equations by hand is essential, but proficiency with computational tools like MATLAB or Python can enhance your ability to handle more complex or numerically intensive problems.

It's important to note that the skills you develop here will be applied extensively in other areas of your studies and future work. Topics like fluid dynamics, thermodynamics, and mechanical design all rely heavily on the principles of differential equations discussed in this chapter. Further details on these applications will be explored in subsequent chapters.

Practice

Cartesian Coordinates Application: Given the equation 3x+4y=12, convert it into the slope-intercept form and sketch the line. Identify its intercepts with the Cartesian axes.

Circle Geometry Problem: Identify the center and radius of the circle that is represented by the given equation $x^2 + y^2 - 6x + 8y = 0$. Sketch the circle and identify any points where it intersects the axes.

Analyzing Quadratic Surfaces: Identify the nature of the surface described by the equation $z = x^2 + y^2$. Describe its shape and possible engineering applications.

3D Analytic Geometry: For the vectors u=⟨1,2,3⟩ and v=⟨4,−5,6⟩, calculate their dot product and discuss the angle between them.

Vector Field Application: Consider a field where the velocity vector at any point (x,y,z) is given by v=⟨y,−x,z⟩. Determine if this field is conservative.

Rate of Change: The function provides the location of a particle $s(t) = t^3 - 6t^2 + 9t$, where t is time in seconds, find the velocity of the particle at t=3 seconds.

Optimization Problem: Identify the dimensions that use the least material for a box with an open top and a square base, which has a volume of 32,000 cm³.

Integration in Mechanics: Measure the work done by a force F=⟨2x,3y,z⟩ moving a particle along the path from (1,0,0) to (1,2,3).

Differential Calculus: Given that $f(x) = x^4 - 4x^3 + 6x^2$, use differential calculus to find the local maxima and minima of f(x).

Integral Calculus: Evaluate the integral $\int_0^\pi \sin \sin(x)\, dx$ and interpret its physical significance in the context of wave motion.

Applying Theorem of Calculus: If F(x) is the antiderivative of $f(x) = 3x^2$ and F(1)=0, find F(3)

Thermal Dynamics Application: If the temperature distribution in a rod is given by $T(x) = 100e^{-0.02x}$, where x is the distance along the rod in cm, calculate the rate of decrease of temperature at x=10 cm.

Fluid Flow Problem: Apply the continuity equation to calculate the velocity of water in a pipe that reduces from 4 cm to 2 cm in diameter, given that the velocity in the larger section is 3 m/s.

Inertia: Determine the moment for a slender rectangular plate with mass m and dimensions a and b, about an axis perpendicular to the plate and passing through its center.

Solving Differential Equations: Solve the differential equation $\frac{dy}{dx} + 3y = 6x$, with the initial condition y(0)=3.

Fourier Series Application: Describe how a Fourier series could be used to model the oscillations of a bridge under periodic load and identify the primary frequencies of oscillation.

Engineering Design Scenario: Design a cam profile described by the equation $r(\theta)=1+0.5\sin(3\theta)$, where r is the radius in meters. Discuss how changes in the frequency component affect the cam's operation.

Complex Problem Solving: Employ the principle of superposition to calculate the cumulative deformation at a specific point in a material subjected to fluctuating loads, assuming the material behaves linearly.

Practical Integration: A force $F(x)=10x$ newtons is applied to a spring. Determin the total energy stored in the spring as it stretches from 0 to 1 meter.

Exam Preparation Strategy: How would you approach a problem involving the optimization of material use in a structural beam? Outline the steps and equations you would use.

Chapter 4: Statistical Strategies for Engineering Success

	Formula	Description	Graphical Representation
Mean (Average)	$\bar{x} = \dfrac{1}{n}\sum_{i=1}^{n} x_i$	The mean is the sum of all data points divided by the number of points.	
Median	Median = $\begin{cases} \left(\dfrac{n+1}{2}\right) \text{th term} & \text{if } n \text{ is odd} \\ \dfrac{\left(\dfrac{n}{2}\right)\text{th term}+\left(\dfrac{n}{2}+1\right)\text{th term}}{2} & \text{if } n \text{ is even} \end{cases}$	The median is the middle value of a dataset.	
Mode	Mode = Most frequent value in the dataset	The mode is the most frequently occurring value in a dataset.	
Standard Deviation (σ)	$\sigma = \sqrt{\dfrac{1}{n}\sum_{i=1}^{n} n(x_i - \bar{x})^2}$	The standard deviation indicates how spread out the values in a dataset are from the mean.	
Variance (σ^2)	$\sigma^2 = \dfrac{1}{n}\sum_{i=1}^{n}(x_i - \bar{x})^2$	The variance is the average of the squared differences from the mean.	
Normal Distribution	$f(x) = \dfrac{1}{\sqrt{2\pi\sigma^2}} e^{-\dfrac{(x-\mu)^2}{2\sigma^2}}$	The normal distribution models continuous data with a symmetric, bell-shaped curve centered around the mean.	
Binomial Distribution	$P(X = k) = \binom{n}{k} p^k (1-p)^{n-k}$	The binomial distribution represents the count of successes within a set number of independent Bernoulli trials.	

Poisson Distribution	$P(X = k) = \frac{\lambda^k e^{-\lambda}}{k!}$	The Poisson distribution describes the frequency of events happening within a specific period of time or area.	
Correlation Coefficient (r)	$r = \frac{\sum(x_i - \bar{x})(y_i - \bar{y})}{\sqrt{\sum(x_i - \bar{x})^2 \sum(y_i - \bar{y})^2}}$	The correlation coefficient quantifies the degree to which two variables are related.	
Regression Analysis	$Y = \beta_0 + \beta_1 X_1 + \beta_2 X_2 + \ldots + \beta_n X_n + \epsilon$	Regression analysis examines how a dependent variable is related to one or more independent variables.	
Multiple Regression	$Y = \beta_0 + \beta_1 X_1 + \beta_2 X_2 + \ldots + \beta_n X_n + \epsilon$	Multiple regression models the relationship between a dependent variable and multiple independent variables.	
Hypothesis Testing	Various formulas depending on the test (e.g., Z-test, t-test)	Hypothesis testing is used to determine if there is enough evidence to reject a null hypothesis.	

Understanding and Applying Key Statistical Measures

The Importance of Statistical Measures

Statistical measures provide engineers with the ability to effectively analyze variability, trends, and probabilities within sets of data. Whether assessing the reliability of components, the consistency of manufacturing processes, or the likelihood of certain operational outcomes, statistics are indispensable.

Central Tendency and Dispersion

Here are some of the key statistical measures that every engineer should master to enhance their data analysis skills and make informed decisions with confidence: mean, median, mode, standard deviation, and variance

1. **Mean (Average):** The mean is calculated by adding all data points together and dividing by the total number of points. It provides a central value that summarizes the entire dataset, crucial for

understanding overall performance metrics such as average material stress or typical energy usage.

$$\text{Mean} = \frac{\sum \text{data points}}{\text{number of data points}}$$

2. **Median:** The median divides your data set into two equal parts. This measure is particularly useful in skewed distributions, as it provides a more robust central tendency that isn't as influenced by outliers or skewed data, typical in manufacturing defect analyses.

   While a specific formula isn't applicable for median, the method involves arranging the data points in order and finding the middle value, or the average of two middle values in an even-numbered dataset.

3. **Mode:** To identify the most common occurrences, such as the most frequent cause of failure in a system It may be helpful to understand the mode.

   The mode is identified by finding the most frequently occurring value in the dataset.

4. **Standard Deviation and Variance:** These measures of dispersion are critical for understanding the spread of your data. It offers a clear view of variability around the mean, crucial for evaluating quality control and risk management in material strength and tolerances.

$$\text{Variance} = \frac{\sum (X - \text{Mean})^2}{n - 1}$$

To find the standard deviation, you simply take the square root of the variance.

Design of Experiments (DOE) in Mechanical Engineering

DOE is an organized approach designed to explore how various factors influencing a process correlate with its results. In mechanical engineering, it is critical to minimize costs and resource use, so optimizing and improving processes, remains a priority. Here's an overview of how DOE is applied in engineering and its critical role in innovation and quality control.

Principles of DOE

DOE involves planning, conducting, analyzing, and interpreting controlled tests to evaluate the effects of various factors on a process outcome. This technique allows engineers to:

- **Identify Critical Factors:** Quickly determine which variables have the most significant impact on product performance.

- **Optimize Processes:** Adjust process parameters to achieve optimal performance, ensuring products meet quality standards and regulatory compliance.
- **Enhance Product Quality:** Systematically test changes to design and process factors to improve the quality and durability of engineering products.

The key components representing these techniques are:

1. **Factorial Design:** This involves testing two or more factors simultaneously, providing insight into the interactive effects of these factors on the output. For mechanical engineering, this could mean exploring different materials and manufacturing techniques to determine the best combination for product durability and cost-efficiency.
2. **Randomized Block Design:** Useful when experiments require controlling variability across batches or runs. For instance, testing a new automotive part design under various environmental conditions to ensure consistency across all tests.
3. **Taguchi Methods:** These focus on robust design and are used to develop products that perform consistently in various operating conditions. By using Taguchi methods, engineers can design products that are less sensitive to variations in environmental factors.

We now'll see the Application and Benefits:

- **Reduced Testing Costs:** By systematically planning experiments, engineers can obtain significant amounts of data from a limited number of tests, reducing resource expenditure.
- **Improved Product Reliability:** DOE helps identify and mitigate potential failure modes in design before products reach the market.
- **Accelerated Innovation:** Fast-tracks the development process by rigorously testing and optimizing designs in a controlled, systematic manner.

Consider the development of a new alloy for aircraft engines, where strength and heat resistance are critical. By applying this systematic method, engineers can experiment with different combinations of alloying elements and heat treatment processes to identify the optimal composition and process settings that deliver the highest performance.

Probability Distributions

We may take this concept for granted but we absolutely would not want to, which is why engineers must be familiar with various probability distributions to model and predict behavior:

- **Normal Distribution:** Used for quality control and assurance, iIt allows engineers to model manufacturing variability and is instrumental in setting tolerance levels, ensuring products meet consistent quality standards.

$$P(x) = \frac{1}{\sigma\sqrt{2\pi}} e^{-\frac{1}{2}\left(\frac{x-\mu}{\sigma}\right)^2}$$

- **Binomial Distribution:** Particularly useful in quality testing scenarios where the outcomes are binary, such as pass/fail conditions. This assists in calculating the likelihood of achieving a specified number of successful outcomes across a series of independent trials, crucial for assessing the quality of products.

$$P(k; n, p) = \binom{n}{k} p^k (1-p)^{n-k}$$

- **Poisson Distribution:** Often applied in reliability engineering and failure analysis, especially for predicting the number of failures or events over a specified period or area.

$$P(k; \lambda) = \frac{\lambda^k e^{-\lambda}}{k!}$$

Probability distributions are only the beginning. To fully exploit the power of statistical analysis, engineers must also delve into correlation and regression analysis

Correlation and Regression Analysis

- **Correlation Coefficients:** These quantify the degree to which two variables are related. Engineers use this measure to examine the relationship between variables, such as load and deformation or temperature and pressure.
- **Regression Analysis:** A statistical tool for modeling and analyzing relationships between variables. It's extensively used to predict a dependent variable using one or more independent variables, such as predicting the stress on a beam based on load and beam dimensions.

$$Y = \beta_0 + \beta_1 X_1 + \beta_2 X_2 + \ldots + \beta_n X_n + \epsilon$$

Where Y is the dependent variable, X_i are independent variables, β_i are coefficients, and ϵ is the error term.

Hypothesis Testing

Hypothesis testing is a powerful statistical technique employed in engineering to make data-driven decisions. This technique involves setting up a hypothesis about a population parameter and then using sample data to support or reject this hypothesis. In the context of mechanical engineering, it is utilized to validate the reliability and performance of new product designs or manufacturing processes. They conduct tests to determine whether parts meet the required standards under specified conditions, such as stress levels or operational environments. This systematic approach ensures that products are reliable and safe for consumer use, reducing risks and optimizing production efficiencies.

Regression Analysis

This statistical tool estimates relationships among variables. This approach enables the prediction and modeling of a dependent variable's behavior using independent variables. For example, by analyzing historical data on material performance under different conditions, engineers can predict how new materials will behave under similar stress. This predictive capability is crucial for optimizing designs, improving material selection, and enhancing operational safety by foreseeing potential failure points due to fatigue under varying loads and environmental conditions.

Advanced Regression Techniques

On the other side we have more complex forms such as multiple regression which is imperative for dealing with the multifaceted nature of engineering problems. Multiple regression allows us to explore the relationship between one dependent variable and two or more independent variables, providing a more comprehensive analysis of the factors that influence engineering outcomes.

1. Multiple Regression: Multiple regression is particularly useful in scenarios where several variables may influence a response variable. For instance the durability of a component could depend on factors such as material hardness, temperature, and exposure to corrosive environments simultaneously. Multiple regression helps in quantifying how much each factor might affect durability, thus allowing engineers to preempt on the choice of materials.

2. Interaction Effects: Another critical aspect of advanced regression techniques is the ability to analyze interaction effects between variables. Effects of interaction occur when the influence of an independent variable on the variable one changes based on the value of another independent one. Properties of materials might change under different temperature and pressure conditions, affecting their overall performance.

3. Polynomial Regression: Polynomial regression extends the linear model to include polynomial terms, which can be used to model the non-linear relationships that are common in engineering data. For example, the stress-strain curve of a material, which often exhibits non-linear behavior, can be modeled more accurately with polynomial regression, providing a better understanding of the material's mechanical properties.

4. Logistic Regression: While primarily used for binary outcomes, logistic regression can be essential when the outcome of an engineering test is pass/fail. This method calculates the likelihood of an event occurring by fitting data to a logistic curve. It is widely used in predictive maintenance and reliability testing in engineering to predict the likelihood of a system failure, which helps in planning maintenance schedules and improving system design.

Correlation Coefficients

Correlation coefficients quantitatively assess the relationship between two or more variables. This metric is critical in engineering for predicting how changes in one factor might influence another, enabling more informed decisions in design processes. For instance, engineers use correlation coefficients to understand how variations in load conditions can affect material deformation. This understanding can significantly impact the selection of materials and the design of structures to ensure safety and functionality under different stress conditions.

Practice

Calculate the Mean: Given a dataset of tensile strengths of a metal rod from 10 samples: [350, 370, 360, 340, 360, 355, 365, 375, 360, 345] psi, calculate the mean tensile strength.

Determine the Median: If the production times for a component are [4.5, 3.2, 4.8, 4.0, 5.1, 3.8, 4.9, 3.7, 4.2, 4.8] hours, find the median production time.

Find the Mode: Given the following failure times in hours [200, 150, 150, 250, 300, 150, 400, 200, 200], identify the mode of the dataset.

Standard Deviation: For the set of values [2, 3, 5, 6, 7, 9], calculate the standard deviation and discuss what it tells about the data spread.

Factorial Design Problem: Design a simple factorial experiment to test the effect of temperature (high, low) and pressure (high, low) on the yield strength of a polymer. What interactions would you be particularly interested in?

Randomized Block Design Application: Explain how you would use a randomized block design to test a new automotive paint for durability under different weather conditions.

Taguchi Methods: How would you apply Taguchi methods to optimize the fuel efficiency of an engine? Consider factors like air-fuel ratio and ignition timing.

Analyze Probability with Normal Distribution: What are the chances that a component, chosen at random and whose lengths follow a normal distribution with an average of 10 cm and a standard deviation of 0.1 cm, exceeds 10.2 cm in length?

Binomial Distribution Scenario: A quality control test for switches shows that 5% are defective. What is the probability that in a sample of 20 switches, exactly two will be defective?

Poisson Distribution Application: If a factory experiences an average of 2 machinery breakdowns per week, what is the probability of having exactly 4 breakdowns in a week?

Correlation Coefficient Calculation: Given data points for temperature and pressure (in units): [(100, 50), (150, 55), (200, 60), (250, 65), (300, 70)], calculate the correlation coefficient and interpret its meaning.

Regression Analysis: Using the data points provided, develop a linear regression model to predict pressure based on temperature. What would be the estimated pressure at 225 units of temperature?

Hypothesis Testing: Suppose you want to test if a new alloy has a higher tensile strength than the industry standard of 500 psi. If your sample of 30 tests yields an average strength of 510 psi with a standard deviation of 15 psi, perform a hypothesis test to determine if the new alloy is significantly stronger at the 0.05 significance level.

Multiple Regression: How would you model the stress experienced by a beam as a function of both beam width and load using multiple regression? Assume you have data for both variables and stress measurements.

Interaction Effects in DOE: Explain how you would analyze interaction effects between material type and temperature in a DOE testing material fatigue.

Polynomial Regression Use: For a dataset that shows the nonlinear relationship between load and displacement, fit a polynomial regression model. What degree of polynomial would you test first?

Logistic Regression: Design a logistic regression analysis to predict the pass/fail outcome of a heat tolerance test for electronic components.

Dimensional Analysis in Testing: For a test measuring the vibration frequency of a new engine type, ensure your dimensional analysis is consistent. What units would you expect your frequency output to be in?

Computational Tools: Which computational tools would you use to solve a complex differential equation derived from your regression analysis? Why?

Practical Application: Determine the z-score for a manufactured shaft that has a diameter of 5.05 inches, given a normal distribution with an average diameter of 5 inches and a standard deviation of 0.02 inches. Explain what this z-score indicates about the manufacturing process.

Chapter 5: Ethical Engineering in Practice

Navigating Professional Ethics and Intellectual Property

Engineering ethics are principles that guide our professional conduct through moral judgements and decisions. Adhering to these ethics is crucial because they foster trust and safety in engineering practices. Central to professional ethics are:

- **Honesty and Transparency**: Always provide truthful and accurate information in your work. This includes acknowledging the limitations of your knowledge and not misleading or deceiving others with incomplete data.
- **Integrity:** Mechanical engineers are required to demonstrate the utmost levels of honesty and integrity. This involves being truthful and transparent in all professional interactions and ensuring that their work is conducted with the utmost accuracy and fairness.
- **Accountability and Responsibility**: Take responsibility for your actions. This means ensuring that your work complies with technical standards and safety requirements, and being accountable for any outcomes.
- **Fairness and Equality**: Treat all colleagues and clients equally, without discrimination. Fairness also means competing ethically and respecting the rights and contributions of others.
- **Respect for Intellectual Property**: A fundamental tenet is respecting the intellectual contributions of others. This includes properly citing the works of other engineers and avoiding plagiarism in all professional activities.
- **Professional Competence**: Engineers should only undertake tasks in areas of their competence. They should also commit to lifelong learning to maintain and enhance their professional skills.

In practice, these codes dictate how engineers interact with clients, manage conflicts of interest, and how they engage in competitive practices. They ensure that all professional activities are conducted with fairness and honesty, contributing to a level playing field in the engineering community.

Intellectual Property: Understanding Your Rights and Responsibilities

Intellectual property (IP) in engineering refers to the creation of the mind that can be legally protected, including inventions, designs, and proprietary processes. As engineers, respecting IP rights involves:

- **Patents**: If you create a new and useful process, machine, or composition of matter, understanding patent laws is useful for safeguarding these inventions. Patents provide you with exclusive rights to your creations, preventing others from making, using, or selling them without your consent. This protection lasts for a specific period, usually 20 years from the patent application filing date, offering a considerable market advantage.

- **Copyrights**: These protect the expression of ideas, such as software code; this can also include software code, engineering drawings, schematics, and technical documentation. Knowing how to protect your work under copyright law ensures that your creative outputs are not used without authorization. Typically lasts for the life of the author plus 70 years.
- **Trade Secrets**: Trade secrets include manufacturing or industrial secrets and commercial secrets. Confidential information must have significant commercial value, such as formulas, practices, processes, designs, instruments, patterns, or compilations of data. Unlike patents, trade secret protection remains in effect indefinitely as long as the secret is not disclosed to the public.

Engineers must implement adequate security measures to protect such secrets, which can include non-disclosure agreements (NDAs) and physical and electronic security measures.Moreover, understanding IP not only protects your creations but also ensures respect for the innovations of others, avoiding legal disputes and fostering a culture of innovation and respect within the engineering community.
- **Industrial Designs**: These rights safeguard the visual design of objects that are not solely functional. An industrial design involves the creation of a shape, configuration, or composition of pattern or color, or a combination of pattern and color in a 3D form that possesses aesthetic value. An industrial design can be a two- or three-dimensional pattern utilized to manufacture a product, industrial good, or handicraft.Integrating Ethics and IP Knowledge into Engineering Practice

Navigating professional ethics and IP requires integrating them into your daily engineering practice. This means:

- Continually educating yourself about ethical standards and legal updates in IP.
- Developing the habit of evaluating the ethical dimensions of your engineering decisions.
- Seeking advice from experienced colleagues or legal experts when in doubt.

Societal Impacts and Responsibilities of Mechanical Engineers

As a mechanical engineer, the impact of my work on society is profound and far-reaching. You need to recognizing the significance of our contributions in ethical practice and the advancement of community welfare. Our engineering solutions can shape the quality of life and the environment, making it essential to consider the broader implications of our decisions.

Environmental Stewardship

One of the paramount responsibilities is environmental stewardship. In my practice, ensuring sustainability in engineering projects is not just about compliance with regulations—it's about leading the way in innovative, eco-friendly solutions. Whether it's developing energy-efficient machines, utilizing renewable energy sources, or minimizing waste, our goal is to protect natural resources for future generations. This approach preserves the environment and also sets a standard for responsible engineering. Here's a few sustainable Engineering Practices:

- **Life Cycle Assessment (LCA)**: It involves evaluating the environmental impacts associated with all the stages of a product's life from raw material extraction through materials processing, manufacture, distribution, use, repair and maintenance, and disposal or recycling. Engineers can use LCA to identify areas where improvements can be made to increase energy efficiency and decrease environmental degradation, ensuring every stage of the process is optimized for minimal environmental impact.

- **Green Manufacturing**: This involves the development and implementation of manufacturing processes that minimize waste and energy consumption. Techniques such as precision engineering and the use of biodegradable materials are explored to ensure that production processes are as environmentally friendly as possible.

- **Renewable Energy Integration**: In mechanical engineering, there is a growing trend towards the integration of renewable energy sources such as solar, wind, and bioenergy into traditional systems.

Safety and Public Welfare

Furthermore, ensuring the safety and well-being of the public is central to our professional responsibilities. Every design and project must undergo rigorous safety assessments to prevent any potential hazards. This involves meticulous risk management and adherence to safety standards to ensure that all engineering outputs are reliable and secure. For instance, when designing machinery or vehicles, the safety of the end-user and the public is the top priority, as You know. Ensuring robust safety measures enhances public trust in engineering innovations.

Advanced Safety Practices:

1. **Predictive Safety Analysis**: Advanced techniques like predictive modeling and simulation identify potential system failures before they occur. This proactive approach allows engineers to address risks at the design stage, rather than reacting to failures after deployment.

2. **Human Factors Engineering**: This field incorporates understanding of human capabilities, limitations, and traits into the design of tools, machines, systems, tasks, jobs, and environments to ensure productive, safe, comfortable, and efficient human use. Including aspects such as ergonomics and user interface design can enhance the safety and usability of engineered products, ensuring they meet the needs of diverse users while minimizing the risk of error and injury.

3. **Safety Protocols and Emergency Response**: Beyond standard safety assessments, developing and implementing comprehensive safety protocols and emergency response strategies are crucial. These protocols should include clear guidelines on how to respond to accidents and safety breaches, training for personnel on safety practices, and regular drills to ensure everyone is prepared for emergency situations.

Social Responsibility

As engineers, our role extends beyond the confines of technical problem-solving; we wield a unique capability to address and ameliorate societal challenges through innovative technology. This immense power to make a significant difference in the world carries with it a profound sense of social responsibility. We are tasked not just with creating and implementing technology but ensuring that these innovations are thoughtfully designed to be accessible and promote equality, ultimately enhancing the quality of life across diverse communities.

Take, for instance, the field of healthcare technology. Our challenge here is to ensure these technologies are affordable and user-friendly, making advanced care accessible even in under-resourced settings. Or think about the transportation sector. Our responsibility extends to crafting systems that are not only efficient but sustainable. This involves designing solutions that integrate renewable energy, reduce emissions, and incorporate smart technology to ease congestion and enhance safety.

Moreover, the increasing severity and frequency of natural disasters due to climate change call for innovative engineering solutions in disaster resilience. From constructing flood defenses to designing earthquake-resistant infrastructure, the technology we develop plays a pivotal role in protecting communities from the catastrophic impacts of natural disasters. Our work ensures that cities can withstand adverse conditions, safeguarding lives and the continuity of communities.

Remember, developing these technologies doesn't involves deepen existing inequalities but rather promote inclusivity. This means designing with an eye toward universal accessibility, ensuring that all members of society—regardless of age, disability, or socioeconomic status—can benefit from our innovations.

In conclusion, the societal impacts of mechanical engineering are significant, encompassing environmental, safety, and social dimensions. As we advance in our careers, integrating these responsibilities into our daily practices is both beneficial and necessary for the long-term sustainability of our profession and the planet.

Detailed Scenario

Managing Crucial Information During Testing

Imagine you are the head of team tasked with developing a new mechanical device, specifically a critical component for automotive safety systems. During testing phases, your team discovers that the device might not function properly in extremely low temperatures, an issue that had not been anticipated and could compromise the vehicle's safety in certain environments.

As the lead engineer, you face a tough choice. On one hand, there is pressure to meet production deadlines and keep costs within budget. On the other, there is an imperative need to ensure product safety.

- Step 1: In-depth Technical Evaluation

You decide to immediately call an emergency meeting with the engineering team to review all the data collected. During the meeting, you collectively analyze the device's performance tests at various temperatures and determine that the issue occurs only below a specific temperature threshold. Various approaches to solve the problem are discussed, including strengthening the materials used or modifying the device design.

- Step 2: Transparent Communication

After evaluating the options, you decide to inform the management and clients about the issue. You prepare a detailed report describing the defect, potential solutions, and the expected impact on production timelines and costs. This document emphasizes the importance of safety and functionality of the product above all other considerations. Reason why you are willing to consider a delay in the schedule rather than propose a project that does not meet appropriate safety standards.

- Step 3: Decision and Implementation

With management's consent, you opt for a solution that might slightly delay production but will ensure that the device is safe and functional. You implement a design review plan and intensify testing to ensure that the problem is resolved before mass production begins.

This approach, based on honesty and transparency, ensures that the final product is safe for use and that customers remain confident in your company's ability to handle issues responsibly. Being proactive in communicating and resolving the problem not only mitigates risks but also strengthens the long-term reputation of your company as a reliable and ethically responsible manufacturer.

Managing Conflicts of Interest with Integrity

Now, suppose you are leading a significant project tasked with purchasing essential materials for an important construction. One of the potential suppliers is owned by a close friend of yours. His company is known for the exceptional quality of materials, but the prices and delivery times are not competitive compared to other suppliers who offer better terms.

Detailed Examination of Proposals: You begin with a thorough analysis of all suppliers' proposals. You organize review sessions with your team to scrutinize each offer, highlighting costs, quality, supply reliability, and alignment with project goals.

Establishing Objective Criteria: You define selection criteria based on objective parameters such as cost, quality, technical compatibility, and supplier reliability. Any decision must be clearly justified and based on concrete facts.

Communication of Decisions: After evaluating all options, you choose the supplier that best meets the project's criteria, even if it means not selecting your friend's company. You explain your decisions to all stakeholders, including the non-selected suppliers, with transparency and detail, referencing the predefined criteria and demonstrating how the final choice is most advantageous for the project.

Maintaining Professional Relationships: Despite opting for another supplier, you commit to maintaining a cordial and professional relationship with your friend's company. This gesture demonstrates professionalism and also strengthens future networking without compromising ethical standards.

Monitoring and Ongoing Evaluation: Once the collaboration with the chosen supplier is initiated, you implement a monitoring and evaluation system to ensure that the supplied materials meet all project specifications and standards. This step ensures that the decision made leads to the desired results and to mitigate any potential issues.

Safeguarding Intellectual Property

In this setting, you're a part of a groundbreaking venture that involves several companies coming together to create an innovative mechanical device. Each company contributes proprietary technical knowledge crucial for the device's development. The foremost challenge in such a collaboration is ensuring that every piece of shared information is secured against unauthorized use and leakage.

Initial Agreement Setting: Before any information exchange begins, you take the lead in organizing a foundational meeting to discuss the terms of engagement. This isn't just about legality but setting the tone for collaboration. You advocate for a collective agreement where all parties commit to respect each other's intellectual contributions.

Custom Security Protocols: Instead of relying solely on standard non-disclosure agreements, your initiative involves developing custom security protocols tailored to the specific nature of the shared information. These might include unique data access levels and specialized encryption methods for different types of shared documents and communications.

Interactive Workshops on IP Sensitivity: You organize a series of interactive workshops for all participating teams. These are designed to not only inform but also engage participants in scenarios that illustrate potential breaches and their consequences.

Continuous Improvement Cycle: You establish an ongoing review cycle that assesses the effectiveness of the implemented security measures. This isn't a once-and-done audit but a dynamic process of continual refinement and adaptation to new security challenges as the project evolves.

Crisis Management Plan: Lastly, you outline a clear and detailed crisis management plan. This plan is immediately activated in the event of any breach or suspicion of unauthorized information use. It details steps for containment, assessment, and rectification, along with predefined communication strategies to manage potential fallout.

The scenarios explored underscore the significant societal impact of mechanical engineering. Ethical decision-making in these instances exemplifies how engineers can navigate complex situations while upholding standards that benefit their projects.

These case studies demonstrate the need to foster a culture of transparency, accountability, and mutual respect in every facet of their work.

The integration of these ethical responsibilities into daily practices is crucial for the long-term sustainability of our profession and the planet. This commitment not only fulfills professional obligations but also drives innovation and societal advancement, reinforcing the pivotal role engineers play in shaping a safe, sustainable future.

Practice

Ethical Dilemma Scenario: You discover a minor calculation error in a project report that could potentially save the client money. Correcting it might delay the project. What do you do?

Intellectual Property Rights: Describe the steps you would take to patent a new mechanical device you invented.

Conflict of Interest: Imagine you are asked to approve a design created by a close relative. How would you handle the situation to maintain fairness and integrity?

Professional Competence: You are assigned a task that is slightly beyond your current area of expertise. How do you proceed to ensure the work meets professional standards?

Handling Sensitive Information: What measures would you implement to protect the trade secrets of a new manufacturing process you developed?

Environmental Stewardship: Propose an engineering solution that could reduce energy consumption in manufacturing facilities. Describe how it aligns with sustainable practices.

Safety Assessment: Outline the steps for conducting a safety assessment for a new consumer product.

Respecting Copyrights: You need to use a copyrighted software tool for a project. What steps do you take to ensure you comply with copyright laws?

Fairness in Practice: How would you ensure all team members have equal opportunities to contribute to a project, especially in a diverse team?

Professional Accountability: Describe a situation where you would need to take responsibility for a project failure and the steps you would take to rectify the situation.

Applying DOE for Quality Improvement: Design a DOE to test different materials for a high-stress component to determine which material performs best under prolonged use.

Advocating for Public Welfare: Propose an engineering project that could significantly improve community welfare. Outline the project's scope, potential impact, and ethical considerations.

Patent Analysis: You find a competitor's product that closely resembles a design your company patented. How do you proceed in assessing potential patent infringement?

Ethical Competitive Practices: Describe how you would handle a situation where you have insider knowledge about a competitor's bid on a project.

Cultural Sensitivity in Global Projects: Discuss how you would manage a project in a different cultural setting to ensure respect for local norms and practices.

Renewable Energy Integration: Propose a mechanical engineering project that utilizes renewable energy sources. Describe the project's benefits to both the environment and the industry.

Human Factors in Safety Design: Using an example, explain how you would integrate human factors engineering into the safety design of a new product.

Innovation and IP: Discuss the importance of IP management in fostering innovation within an engineering team.

Ethics in Autonomous Systems: Consider the ethical implications of developing autonomous mechanical systems. What ethical guidelines would you establish?

Sustainability and Lifecycle Assessment: How would you conduct a lifecycle assessment for a new product, and what factors would you consider to enhance its sustainability?

Chapter 6: Economics of Engineering Decisions

Time Value of Money and Cost Analysis in Engineering Projects

This principle posits that a dollar today is worth more than a dollar in the future due to its potential earning capacity. This core concept is used to evaluate investment opportunities within engineering projects, ensuring that decisions are not only technically sound but also financially prudent.

To begin, let's explore the basic mechanisms of TVM. It includes calculations such as present value (PV), future value (FV), net present value (NPV) and internal rate of return (IRR) are calculations that assist engineers in assessing whether a project is likely to generate a positive return, taking into account variables such as opportunity cost, risk and inflation.

1. **Present Value (PV)**: It involves calculating the present value of future cash flows to establish their worth today. This is crucial in evaluating the viability of projects with long-term benefits or costs. For instance, determining the PV of expected future revenue from a new technology helps in making the investment decision today.

$$PV = \frac{FV}{(1+r)^n}$$

2. **Future Value (FV)**: Conversely, FV estimates the worth of an investment at a future time, assuming a predetermined rate of return. This calculation is used when planning for future expenses or savings, such as the anticipated cost of upgrading plant equipment five years from now, allowing engineers to set aside adequate resources.

$$FV = PV \times (1+r)^n$$

3. **Net Present Value (NPV)**: This is arguably the most crucial of all time value of money calculations for project evaluation. NPV involves calculating the total present values of all cash inflows and outflows over the lifespan of a project. A positive one indicates that the project is likely to generate a value greater than its cost, making it a financially viable option. It's particularly useful in comparing multiple project options where each has differing cash flow and timelines.

$$NPV = \sum_{t=1}^{n} \frac{C_t}{(1+r)^t}$$

4. **Internal Rate of Return (IRR)**: This calculation is used to determine the profitability of potential investments. IRR is the discount rate that makes the NPV of all cash flows from a particular project equal to zero. In simpler terms, it provides a clear benchmark to gauge the efficiency of the investment compared to the firm's required rate of return or other available investment opportunities.

$$0 = \sum_{t=0}^{n} \frac{C_t}{(1+IRR)^t}$$

Why are they so important?

For example, changing the discount rate to more accurately represent the project's risk profile or updating cash flow estimates based on recent market research can yield more precise and relevant financial insights. This adjustment processes are crucial in dynamic industries where market shifts in market conditions and technological progress can substantially affect project scopes and results.

Understanding and applying TVM principles enables engineers to not only predict financial outcomes but also to adapt project plans responsively. Sensitivity analysis using TVM can show how changes in market interest rates or project delays affect the overall viability of a project. This can guide engineers in making preemptive adjustments to project management strategies, ensuring that investments remain robust against financial uncertainties.

TVM calculations are also integral in long-term strategic planning, especially for projects that span multiple decades, such as infrastructure or energy developments. Engineers can use these tools to forecast long-term financial sustainability and assess the impact of various financing options.

Utilizing concepts like NPV and IRR, You can rank projects according to their financial returns and risk profiles. This optimization of resource allocation ensures that capital is invested in projects that are most likely to achieve the desired financial and technical outcomes, maximizing efficiency and shareholder value. It also prevents financial misrepresentation and ensure that projects are conducted transparently and ethically.

Additionally, engineers often engage in cost-benefit analysis (CBA), a technique that contrasts the costs of a project or decision with the benefits it will deliver. This analysis helps quantify the economic value of projects in terms of their potential to increase efficiency, reduce waste, or improve productivity

Project Financing Options for Engineering Projects

Understanding the range of financing options available for engineering projects is essential for making well-informed decisions that meet project requirements and financial strategies. As we have explored the principles of TVM and cost analysis, it's important to keep in mind how different financing methods can impact the overall cost and profitability of projects. Here, we'll delve into the various methods of financing, such as loans, bonds, leases, and public funding, and discuss the implications of each on an engineering project's financial health.

1. Loans Loans are a common financing option for engineering projects, providing immediate capital with the agreement of repayment over time plus interest. The amount paid of the loan is related by the interest rate, which can be fixed or variable. Engineers need to calculate the future financial burden of the loan repayments, utilizing TVM calculations to understand the true cost of the loan over its duration.

2. Bonds: Issuing bonds is another method, particularly for large-scale public or private projects. Bonds are essentially promises to pay holders an amount of interest over a period and then return the principal on the maturity date. the interest rate on bonds, which often mirrors the level of risk associated with the project influences the cost of capital for a project . Understanding how to evaluate the impact of bond financing on project costs is essential for long-term financial planning.

3. Leases Leasing is an option for acquiring equipment or facilities without the upfront costs associated with purchasing. For engineering projects, leases can be structured as operating leases or finance leases, each with different implications for the project's balance sheet and overall costs. To decide whether leasing is more cost-effective than purchasing, engineers need to evaluate the entire cost of leasing, considering commitments and potential options to buy.

4. Public Funding Projects that provide public benefits, such as infrastructure or renewable energy projects, may qualify for public funding. This funding can come in the form of grants, tax incentives, or direct investment. Each funding type has specific criteria and implications for project cash flow and profitability.

Capital Budgeting: Essential Techniques for Engineering Project Valuation

Capital budgeting is a key process in engineering project management, vital for evaluating the financial viability and strategic importance of potential projects. It encompasses various techniques that enable project managers and engineers to assess investment decisions with a thorough understanding of their possible returns and risks. Given its significance in ensuring that resources are allocated efficiently and strategically, capital budgeting should be integrated into the financial analysis framework we've discussed in the Previous section.

Key Techniques in Capital Budgeting

1. **Payback Period**: This is the basic capital budgeting methods, which calculates the time to generate cash flows sufficient to recover the start cost. This metric ensures a quick assessment of liquidity and risk for engineering projects, but provides a failure considering the time value of money, which is essential for accurately comprehending true project costs and benefits.

$$\text{Payback Period} = \frac{\text{Initial Investment}}{\text{Annual Cash Inflows}}$$

2. **Discounted Payback Period:** Enhancing the basic payback model, the discounted payback period integrates the time value of money, an essential concept we've emphasized. By applying a relevant discount rate to cash flows, this method provides a more accurate measure of the time needed for a project to reach profitability in terms of present value. This method is in line with our discussions on NPV and IRR, ensuring a consistent approach to evaluating financial results.

3. **Profitability Index (PI)**: The profitability index, calculated by dividing the present value of future cash flows by the initial investment, serves as a beneficial metric. A PI exceeding 1 suggests that the project is financially beneficial. This index enhances the NPV method by showing the ratio of received to invested value.

$$PI = \frac{\text{Present Value of Future Cash Flows}}{\text{Initial Investment}}$$

Importance of Risk Assessment in Engineering

Risk assessment is integral to engineering because it helps prevent cost overruns, delays, and failures in project execution. To identify it early, means that You can devise strategies to mitigate them effectively, enhancing project reliability and success.

Risk is compost by these main phases:

Risk Identification: The first step is simple called as identify potential risks. This covers a thorough analysis of the project scope, resources, environment, and other relevant factors. Techniques such as brainstorming sessions, Delphi techniques, and SWOT analysis are commonly used to gather comprehensive risk data.

Risk Evaluation: After identifying risks, the next step is to assess their potential impact and the likelihood of their occurrence. This assessment helps prioritize risks according to their severity and the probability of affecting the project. Common tools used in this phase include risk matrices and qualitative and quantitative risk assessment methods.

Risk Mitigation Strategies: After evaluating the risks, appropriate mitigation strategies are formulated. These strategies may include:

- **Avoidance**: Modifying project parameters to mitigate risks.
- **Reduction**: Adopting actions to decrease the impact or likelihood of risks.
- **Transfer**: Transferring the risk to another party, such as via insurance or outsourcing.
- **Acceptance**: Recognizing the risk and planning to manage its effects if it materializes.

Advanced Techniques in Risk Management

- **Monte Carlo Simulations**: This statistical technique uses probability distributions to model and assess the impact of risk on project outcomes. By simulating thousands of scenarios, Monte Carlo simulations provide a detailed analysis of potential outcomes and their probabilities.
- **Sensitivity Analysis**: This method investigates the impact of various independent variable values on a specific dependent one, assuming certain conditions. Sensitivity analysis is crucial in understanding which risks have the most impact on project outcomes and where focus should be prioritized.

Integrating Risk Management with Project Lifecycle

Effective risk management is a dynamic component that is woven through every phase of the project lifecycle, ensuring that risks are identified, assessed, and mitigated from project initiation to completion. This approach aligns with the broader strategies of risk assessment and management previously outlined, which utilize methods like Monte Carlo simulations and sensitivity analysis to forecast and prepare for potential outcomes.

1. **Project Initiation**: At this stage, risk management involves preliminary risk identification and analysis. Potential risks related to project scope, resources, stakeholder expectations, and external factors are cataloged. Techniques such as brainstorming sessions and expert interviews are instrumental here. The integration of Monte Carlo simulations can begin by estimating the probabilities of different risks and their impacts on project timelines and budgets.

2. **Planning Phase**: During planning, detailed risk assessments are conducted. This includes developing risk mitigation strategies and incorporating them into the project plan. Sensitivity analysis is particularly useful at this stage to understand how changes in project variables—such as delays in delivery of materials or fluctuations in labor costs—could affect project outcomes. The results help in refining project strategies, setting realistic milestones, and preparing contingency plans.

3. **Execution Phase**: As the project progresses, continuous risk monitoring and control are critical. The risk management plan is implemented, and ongoing monitoring helps detect new risks as they arise. Techniques like Earned Value Management (EVM) are applied to monitor project performance and its alignment with the risk-adjusted project plan. Adjustments are made as needed based on real-time data and simulation updates, ensuring that the project remains on track and within the defined risk tolerance levels.

4. **Closure Phase**: In the project closure phase, a final risk review is conducted to assess how risks were handled and to document lessons learned for future projects. This retrospective analysis enhances the organization's risk management capabilities and contributes to a refined risk assessment template for future projects.

Advanced Applications of Cost-Benefit Analysis

Let's look at some advanced risk design techniques:

1. **Scenario Analysis**: CBA can be extended to include scenario analysis, where engineers examine various possible outcomes of a project by altering key variables within this model. This could involve assessing the impacts of unexpected changes in material costs, labor rates, or technological advancements. By considering different scenarios, engineers can develop strategies that are robust under various future conditions, ensuring projects are resilient against a range of economic and technical uncertainties.

2. **Environmental and Social Considerations**: Modern CBA includes not only traditional financial metrics but also environmental and social impacts. This involves quantifying benefits like reduced carbon emissions, improved public health, or enhanced community welfare, which may not have direct monetary values but are crucial for sustainable development in order to respect previously points expanded on the ethic section.

3. **Long-term Economic Impacts**: Engineers expand CBA to look beyond immediate project outcomes by incorporating long-term economic impacts such as the enhancement of local economies, the potential for job creation, and the long-term operational costs and benefits. This

broader perspective helps in making informed decisions that account for the enduring impacts of engineering projects on both local and global scales.

4. **Regulatory Compliance and Incentive Analysis**: CBA also helps in understanding the financial implications of regulatory compliance and potential incentives. Engineers can analyze how adherence to new regulations or qualification for governmental incentives can affect the overall cost-benefit ratio of a project, influencing project structure and financing decisions.

5. **Stakeholder Impact Analysis**: Effective CBA must consider the impacts on various stakeholders, including customers, local communities, investors, and governmental bodies. This evaluation guarantees the consideration of the interests of all key stakeholders in the decision-making process for the project, fostering transparency and involvement, as discussed in the previous concepts.

By integrating these financial tools into daily practice, engineers can provide more value to their projects and organizations. They ensure that projects are not only engineered to high standards but also align with strategic financial objectives. This holistic approach to project management ensures sustainability and profitability, cementing the critical role of engineers in strategic decision-making within businesses.

Practice

Calculate Present Value: Calculate the current worth of receiving $10,000 five years in the future, with an annual discount rate set at 5%.

Future Value Calculation: If you invest $15,000 at an annual interest rate of 8%, what will the future value be in 10 years?

Net Present Value Scenario: Determine the net present value (NPV) for a venture that requires an upfront investment of $50,000 and is expected to generate $15,000 each year over a five-year period, using a 6% discount rate. Assess whether this investment is financially worthwhile.

Internal Rate of Return Problem: Determine the IRR for a project with an upfront cost of $100,000 that delivers annual returns of $30,000 for a period of five years.

Loan Amortization Question: If a company takes a $200,000 loan at a 10% annual interest rate to be paid back over 10 years, calculate the annual payment using the annuity formula.

Evaluate Bond Financing: Describe the impact of issuing a bond with a 5% interest rate on the project's cost of capital compared to a bank loan with a 7% interest rate.

Leasing vs. Buying Decision: Calculate the total cost of leasing equipment for 5 years with annual payments of $10,000 versus buying it for $40,000. Assume a discount rate of 5%.

Public Funding Impact: Examine the impact of a $100,000 grant on the net present value (NPV) and internal rate of return (IRR) for a project that requires an initial investment of $500,000 and is expected to generate annual returns of $70,000 over a span of 8 years.

Payback Period Calculation: Determine the time required to recover an initial outlay of $150,000 from annual returns of $50,000.

Discounted Payback Period: Compute the time it takes to recoup an initial investment of $150,000 with yearly inflows of $50,000, discounting at a rate of 6%.

Profitability Index: If a project has a present value of future cash flows of $600,000 and an initial investment of $550,000, what is the profitability index?

Sensitivity Analysis: Assess the impact on the NPV of a project requiring $200,000 initially and yielding $50,000 yearly over five years if the discount rate shifts from 6% to 8%.

Monte Carlo Simulation Usage: Explain how you would use Monte Carlo simulations to assess risk in a project with uncertain cash flows.

Risk Identification Exercise: List potential risks for a new manufacturing facility project and suggest methods to mitigate them.

Environmental Impact CBA: Conduct a cost-benefit analysis for implementing an environmentally friendly technology that costs $100,000 but reduces pollution costs by $20,000 annually.

Stakeholder Impact Analysis: Discuss how the construction of a new plant would affect local stakeholders and how you would address their concerns in project planning.

Capital Budgeting Decision: Compare two potential projects based on their NPV and IRR. Project A involves an initial outlay of $250,000 and provides annual revenues of $60,000 over a six-year period. Conversely, Project B necessitates an investment of $180,000 and yields annual returns of $50,000 for a duration of five years.

Scenario Analysis: Provide a scenario analysis for an engineering project where market conditions might fluctuate significantly. Analyze two scenarios: best-case and worst-case.

Regulatory Compliance Cost Analysis: Estimate the financial impact of new environmental regulations on a project that is expected to cost $1 million.

Advanced Safety Practice Cost Justification: Justify the cost of integrating advanced safety measures into a project, considering both the direct and indirect benefits.

Chapter 7: Mastering the Essentials of Electricity and Magnetism

	Formula
Voltage (V)	V=I×R
Current (I)	$I = \frac{V}{R}$
Resistance (R)	$R = \frac{V}{I}$
Ohm's Law	V=I×R
Kirchhoff's Current Law (KCL)	$\sum I_{in} = \sum I_{out}$
Kirchhoff's Voltage Law (KVL)	ΣV=0 (around a closed loop)
Gauss's Law for Electricity	$\Phi_E = \frac{Q}{\epsilon_0}$
Gauss's Law for Magnetism	ΦB=0
Faraday's Law of Induction	$EMF = -\frac{d\Phi_B}{dt}$
Ampere's Law with Maxwell's Addition	$\oint \vec{B} \cdot d\vec{l} = \mu_0(I_{enc} + \epsilon_0 \frac{d\Phi_E}{dt})$

Fundamental Electrical Principles and Their Applications in Mechanical Engineering

Electricity and magnetism are cornerstones of modern engineering disciplines. Electrical principles such as voltage, current, and resistance form the basis of electrical circuit theory, which is essential for designing and analyzing everything from household appliances to sophisticated industrial machinery.

- Voltage, or electric potential difference, is the driving force that pushes electric charges through a circuit. Usually it is comparable to pressure in hydraulic systems and is critical for determining how much energy is carried by an electric current.

- Current, measured in amperes, represents the flow of electric charges and is analogous to the flow rate in fluid dynamics.
- Resistance, measured in ohms, quantifies the opposition within the circuit to the flow of current, akin to friction in mechanical systems.

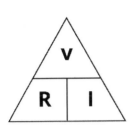

Here's a quick and easy way to definitively learn the formulas for these fundamentals. All you need to do is mentally take a picture of this triangle and cover the variable you need to always have the formulas handy. Let's say you need to find the voltage. Cover V, and what will be left is IxR. Now let's do the reverse example. You need the resistance: cover R, and what will be left is V/I. It's a simple trick that can always come in handy.

These electrical properties are not just theoretical constructs but are extensively applied in the analysis of circuits, from simple household appliances to complex industrial machinery. In household appliances they allow engineers to design efficient and safe electrical systems that can perform specific functions while minimizing energy wastage. For industrial machinery, they handle higher power loads and operate continuously under demanding conditions.

Moreover, the application of these principles extends into the design and troubleshooting of electronic equipment, where the correct application of Ohm's Law and Kirchhoff's Laws helps in predicting the behavior of the circuit under various conditions. These laws make it possible to systematically analyze complex circuits, solve for unknown variables, optimize designs, and enhance functionality while ensuring safety and compliance with technical standards.

Now that we've discussed these, let's consider them appropriately:

- **Ohm's Law**, called by Georg Simon Ohm, is a principal formula within electrical engineering that establishes a straightforward relationship between voltage (V), current (I), and resistance (R). This law is represented as V = I×R, indicating that the voltage across a conductor is directly proportional to the current passing through it, with the proportionality constant being the resistance. Ohm's Law is pivotal for predicting how electrical circuits behave under different electrical charges and resistances, offering a clear and calculable method to assess circuit parameters.
- **Kirchhoff's Laws** further extend the analysis capability in circuits by detailing two essential rules: Kirchhoff's Current Law (KCL) and Kirchhoff's Voltage Law (KVL).
  - **Kirchhoff's Current Law** asserts that the total current entering a junction (or node) must equal the total current exiting the node. This law is founded on the principle of conservation of charge and is crucial for analyzing complex circuits where multiple pathways and junctions exist, allowing for the detailed tracking of current flow in various circuit paths.

$$\sum I_{in} = \sum I_{out}$$

  - **Kirchhoff's Voltage Law** asserts that the sum of all voltages around any closed loop in a circuit must equal zero. This is derived from the conservation of energy principle and is vital for ensuring that the applied voltages in a circuit are balanced by the voltage drops across different components, which is critical for the functional and safe operation of electrical systems.

$$\sum V = 0$$

Together, these laws simplify complex circuit analysis and they ensure the practical application of electrical theories in designing systems that integrate various electrical components efficiently. They allow for the meticulous crafting of circuit designs that meet specific operational standards and safety requirements, crucial for both simple household systems and complex industrial setups.

Electromagnetism Basics

To deepen our understanding, let's explore the foundational Maxwell equations, which describe all electromagnetic phenomena and are indispensable for advanced applications involving electromagnetic fields.

Maxwell's Equations: Maxwell's equations comprise four partial differential equations that serve as the theoretical foundation for classical electrodynamics, optics, and electric circuits. They explain how electric and magnetic fields are produced and modified by charges, currents, and variations in the fields themselves. Here's a brief overview of each:

1. **Gauss's Law for Electricity** states that the electric flux exiting a closed surface is proportional to the charge enclosed by the surface. This law allows you to understand how charges produce electric fields.

$$\oint_S \vec{E} \cdot d\vec{A} = \frac{Q_{enc}}{\epsilon_0}$$

2. **Gauss's Law for Magnetism** reveals that magnetic monopoles do not exist; thus, the total magnetic flux through a closed surface is zero. This is essential for the concept of magnetic field lines always forming closed loops.

$$\oint_S \vec{B} \cdot d\vec{A} = 0$$

3. **Faraday's Law of Induction** shows that a changing magnetic field within a loop of wire induces an electromotive force (EMF) in the wire. This principle is the basis of electric generators and induction motors.

$$\oint_C \vec{E} \cdot d\vec{l} = -\frac{d}{dt} \int_S \vec{B} \cdot d\vec{A}$$

4. **Ampere's Law with Maxwell's Addition** (the Maxwell-Ampere equation) indicates that magnetic fields can be generated in two ways: by electric current (conventional Ampere's Law) and by changing electric fields (Maxwell's addition). This law is pivotal in explaining how capacitors work in AC circuits.

$$\oint_C \vec{B} \cdot d\vec{l} = \mu_0 I_{enc} + \mu_0 \epsilon_0 \frac{d}{dt} \int_S \vec{E} \cdot d\vec{A}$$

Integrating Maxwell's Equations with Mechanical Engineering

Designing Electromagnetic Systems: From magnetic bearings to advanced propulsion systems, Maxwell's equations help engineers design components that utilize electromagnetic fields for motion and force manipulation without direct contact, reducing wear and increasing system lifespan.

- **Electromagnetic Compatibility (EMC)**: Engineers must ensure that electronic and electromechanical devices operate without interfering with one another. Maxwell's equations are critical for predicting and mitigating electromagnetic interference (EMI), ensuring that devices meet regulatory standards.
- **Innovative Applications**: The principles derived from these equations enable the exploration of new technologies such as wireless power transfer and advanced magnetic resonance imaging (MRI) systems.

Electromagnetic Applications in Mechanical Engineering

Electrical principles, foundational to the operation and analysis of myriad systems, extend profoundly into mechanical engineering. Here, these principles are not only essential for power management and machinery operation but also for innovative applications such as in the design of magnetic bearings.

Electromagnetism and Magnetic Bearings

Magnetic bearings represent a prime example of applying electromagnetism in mechanical engineering. By suspending a rotor in mid-air using magnetic fields, these bearings avoid physical contact between parts, thus reducing wear and eliminating the need for lubrication. This setup is particularly beneficial in scenarios where maintenance is challenging or where contamination from lubricants must be avoided, such as in cleanroom environments or in aerospace applications.

Bridging Mechanics and Electromagnetism

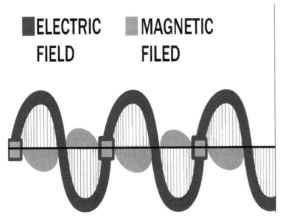

1. **Reduced Mechanical Wear:** Magnetic bearings significantly decrease the wear and tear on components, which enhances the durability and longevity of machinery.
2. **Energy Efficiency:** By eliminating friction, magnetic bearings reduce the energy required for machine operation, contributing to more energy-efficient system designs.
3. **High-Speed Operation:** These bearings can handle very high rotational speeds more efficiently than traditional bearings, which is essential for applications like turbines and high-speed compressors.

Integration with Thermodynamics and Fluid Mechanics

In addition to their mechanical applications, electrical principles are deeply woven into thermodynamics and fluid mechanics. The control and movement of fluids in various engineering systems, from HVAC in buildings to propulsion systems in aerospace, often rely on electrical systems to operate pumps, compressors, and other fluid-moving components. Electromagnetic principles help in optimizing these systems for better performance and energy conservation.

Material Science Applications

Electrical properties also take center stage in material science, They help in developing smarter materials that can interact with electrical systems in innovative ways. For instance, materials with tailored electromagnetic properties can be designed to improve sensing capabilities or enhance the electromagnetic compatibility of different devices.

Detailed Problem-Solving in DC and AC Circuit Analysis

Fundamental Concepts in DC Circuit Analysis

For mechanical engineers, this often relates to the control systems and power distribution aspects of machinery and mechanical devices. Here are the fundamental concepts tailored to the needs of the FE Mechanical exam:

1. **Network Theorems**: Essential for simplifying complex circuits into manageable components, the primary network theorems include:
    - **Thevenin's Theorem**: Simplifies a network with voltage and current sources and resistors to a simple two-terminal circuit consisting of a single voltage source and series resistance.
    - **Norton's Theorem**: Similar to Thevenin's, but provides a simplified circuit with a current source and parallel resistance.
2. **Superposition Theorem**: This theorem analyzes circuits with multiple sources (voltage or current). It involves solving the circuit multiple times, each with only one independent source active, and then superimposing the effects to find the overall solution.
3. **Maximum Power Transfer**: This concept ensures that systems deliver maximum efficiency. It states that maximum power is transferred when the load resistance equals the Thevenin resistance of the circuit supplying power.

Circuit Response to Transient Conditions

In mechanical systems, especially in controls and automation, circuits respond to sudden changes in their operating conditions—known as transient analysis—

This includes studying:

- **RC Circuits**: The response of a resistor-capacitor (RC) circuit to a step voltage input, important for understanding timing circuits.
- **RL Circuits**: The behavior of a resistor-inductor (RL) circuit during transient conditions helps in understanding inductive loads in motor starters and other similar applications.

Practical Implementation of DC Analysis

To bridge these theoretical concepts to real-world applications, consider their implications in the design and troubleshooting of electromechanical systems:

- **Energy Conservation Systems**: Applying maximum power transfer and efficiency calculations to optimize the energy usage in mechanical systems.
- **Automated Control Systems**: Utilizing transient response analysis in the design of control logic for automated systems to ensure stability and responsiveness.

Alternating Current (AC) Circuit Analysis

AC circuits, where the voltage and current vary sinusoidally with time, are more complex due to the presence of capacitors and inductors whose behavior changes with frequency. The analysis of these circuits often involves:

1. **Phasor Diagrams**: Representing sinusoidal functions as rotating vectors (phasors), which simplifies the calculation of voltages and currents in AC circuits.
2. **Impedance (Z)**: A complex number that extends the concept of resistance to AC circuits, encompassing resistance (R), inductive reactance (XL), and capacitive reactance (XC). The total impedance in AC circuits is given by

   $Z = R + j(X_L - X_C)$, where j is the imaginary unit.

3. **Resonance**: Occurs in AC circuits when the inductive and capacitive reactances cancel each other out, leading to maximum current flow at a specific frequency, known as the resonant frequency.

Moreover, understanding the power in AC circuits, which includes concepts of real power (P), reactive power (Q), and apparent power (S), is vital. The relationship among these is expressed through the power triangle, linking them via $S^2 = P^2 + Q^2$

which helps in designing more efficient power systems with better power factor management.

Let's consider in a little more detail

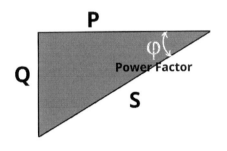

Real power, measured in watts (W), represents the power actually consumed by resistive components to perform useful work. For example, consider a motor. This is the power consumed to keep the motor running efficiently. Reactive power, measured in volt-amperes reactive (VAR), represents the power stored and released by inductive and capacitive components, which does not perform useful work but is necessary for maintaining voltage levels in the system. We could think of this energy as the effort used to keep the spring compressed or stretched without doing actual work, similar to how a swinging pendulum stores energy in motion but doesn't do useful work during its swings. Apparent power, measured in volt-amperes (VA), is the combination of real and reactive power and represents the total power flowing in the circuit. You can think of this as the total effort required to operate a system. For example, if you imagine pushing a box across a rough floor, the apparent power would be the total energy you exert, which includes both the energy that actually moves the box (real power) and the energy used to overcome friction (reactive power).

The equation illustrates how apparent power is the vector sum of real and reactive power, forming a right-angled triangle where the hypotenuse is the apparent power, and the adjacent and opposite sides are the real and reactive powers, respectively.

Understanding the relationship between real power, reactive power, and apparent power is fundamental for designing efficient power systems with effective power factor management. The power factor, defined as the ratio of real power to apparent power (Power Factor = P/S), is a key parameter for evaluating the efficiency of power usage. Essentially, a higher power factor indicates a more efficient system, as it signifies that a greater proportion of the supplied power is being utilized to perform useful work.

Let's consider an example to make this clearer. Imagine you have a factory with numerous machines. The real power is the energy used to run these machines, making them produce goods. The reactive power is like the energy needed to maintain the magnetic fields in the motors, essential for their operation but not contributing directly to production. The apparent power is the total energy the factory draws from the grid to keep everything running.

Engineers aim to improve the power factor to make the system more efficient. A high power factor means that most of the energy drawn is used for productive work. However, if the power factor is low, it means a significant portion of the energy is being wasted on maintaining the voltage levels, leading to inefficiencies.

To improve the power factor, engineers use power factor correction techniques. This often involves adding capacitors or inductors to the circuit. Capacitors provide leading reactive power, which can offset the lagging reactive power caused by inductive loads like motors. This balance reduces the total reactive power in the system, thus increasing the power factor.

For instance, adding capacitors to a motor circuit can reduce the amount of reactive power needed from the power source. This means the S is reduced without changing to the P, leading to a higher power factor. The result is a more efficient system that uses less energy for the same amount of useful work, reducing energy losses and lowering electricity costs.

Practice

Calculate the voltage across a resistor: Given a circuit with a 12-volt battery connected in series with a resistor of 4 ohms, calculate the current flowing through the circuit using Ohm's Law.

Determine the total resistance in a circuit: What is the overall resistance in a circuit where resistors of 2 ohms, 3 ohms, and 5 ohms are connected in parallel?

Analyze current flow in a complex network: Using Kirchhoff's Current Law, determine the missing current in a junction where three currents enter: 5 A, 3 A, and 2 A, and two currents exit: 4 A and 1 A.

Voltage drop across components in a series circuit: Using Kirchhoff's Voltage Law, determine the voltage drop across a 2-ohm resistor and a 3-ohm resistor in a series circuit with a 12 V battery.

Application of Faraday's Law of Induction: Describe how a change in magnetic field would induce an electromotive force in a coil of wire. Provide an example with a hypothetical change in magnetic field over time.

Calculating impedance in an AC circuit: If an AC circuit contains a 4-ohm resistor, a 3-ohm inductive reactance, and a 2-ohm capacitive reactance, calculate the total impedance of the circuit.

Phasor diagram creation: Sketch a phasor diagram for a simple AC circuit with a resistor (R=4 ohms) and an inductor (L=0.5 H) operating at 50 Hz.

Maxwell's Equations application: Describe how to apply Gauss's Law for Electricity to measure the electric field surrounding a point charge that possesses 5 Coulombs.

Transient response in an RC circuit: Calculate the time constant for an RC circuit with a 1 μF capacitor and a 10-k ohm resistor. Determine the duration required for the capacitor's voltage to achieve 63% of its peak following an initial step voltage application.

Use of Thevenin's Theorem in circuit simplification: Simplify a circuit using Thevenin's Theorem where a 12 V battery is connected in series with a 6-ohm resistor and parallel with a 4-ohm resistor.

Power calculation in AC circuits: Determine the actual power used in a circuit with an rms voltage of 120 V and an rms current of 10 A, given a phase angle of 30 degrees between the current and voltage.

Resonance frequency determination: Determine the resonant frequency of an LC circuit with a 0.01 H inductor and a 100 μF capacitor.

Magnetic field calculation using Ampere's Law: Determine the internal magnetic field of a lengthy solenoid that has a current of 2 A and 1000 turns, with a total length of 0.5 meters.

Exploring electromagnetic compatibility (EMC): Discuss how you would test a device for electromagnetic compatibility. What factors would you consider?

Engineering application of magnetic bearings: Describe an industrial application where magnetic bearings could be used to enhance system efficiency and provide a detailed explanation of the operational principles involved.

Safety assessment using electromagnetic principles: Propose a method to ensure safety in the operation of an electrical device using principles of electromagnetism.

Design challenge involving Maxwell's Equations: Design a simple electromagnetic system (e.g., an electromagnet) that could be used to move metallic objects. Use Maxwell's Equations to justify your design choices.

Real-world problem-solving with Kirchhoff's Laws: Given a real-world problem of a malfunctioning traffic light system, explain how you would use Kirchhoff's Laws to diagnose the issue in the circuitry.

Capacitive reactance calculation in an AC circuit: For an AC circuit operating at 60 Hz with a capacitor of 50 µF, calculate the capacitive reactance.

Advanced application of electromagnetism in new technologies: Discuss how electromagnetism principles can be applied to develop wireless charging technology for electric vehicles, including an outline of the technical challenges and potential solutions.

Chapter 8: Statics and Dynamics: The Mechanics of Movement

	Formula
Force Equilibrium	$\sum \vec{F} = 0$
Moment Equilibrium	$\sum \vec{M} = 0$
Newton's First Law	An object will remain stationary unless influenced by an external net force.
Free-Body Diagram (FBD)	Not a formula, but a method to visualize forces and moments.
Moment (Torque)	$\vec{M} = \vec{F} \times \vec{d}$
Couples	$\vec{C} = \vec{F} \times \vec{d}$
Planar Motion - Translational	$\vec{F} = m\vec{a}$
Planar Motion - Rotational	$\vec{\tau} = I\vec{\alpha}$
Angular Velocity	$\omega = \frac{d\theta}{dt}$
Angular Acceleration	$\alpha = \frac{d\omega}{dt}$
Moment of Inertia	$I = \int r^2 \, dm$

Comprehensive Coverage of Statics from Basics to Complex Applications

Statics, a core branch of mechanics, focuses on the study of loads (force and torque, or "moment") on physical systems in static equilibrium. This means that the relative positions of subsystems remain constant over time, or components and structures are at rest under external forces.

In exploring statics, we start with the simple yet profound concepts of force and moment. Forces are vector quantities having both magnitude and direction, essential in the analysis of any physical structure or system. Moments, also vectors, measure the tendency of a force to rotate an object about an axis.

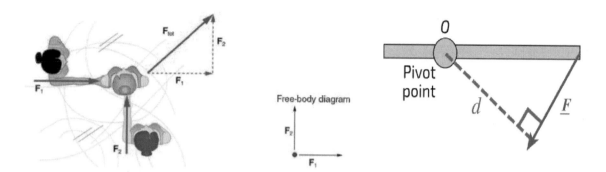

To start we can say that statics is governed by certain essential principles and laws. The most fundamental of these is Newton's First Law of Motion, which states that an object remains at rest unless an external force acts on it.

The state just described is called equilibrium and occurs when the sum of all forces and moments acting on a body is zero. This can be expressed mathematically as: $\sum F=0$ where F represents the forces and M the moments.

The vectors and forces we have just mentioned would be difficult to learn if they could not be represented graphically. That is why this tool we introduce will always accompany you. Having a representation of a concept is always better. It helps us understand and, above all, allows us to give our own interpretation and fix the concept more firmly.

Free-Body Diagrams (FBDs)

An FBD is a simplified representation of a system, isolating a single body and illustrating all external forces and moments acting upon it. By analyzing an FBD, you can apply the equilibrium equations to solve for unknown forces or moments.

To construct an FBD:

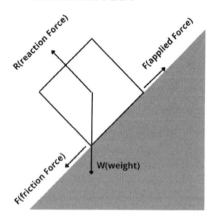

1. **Isolate the body**: Imagine the body cut free from its surroundings.
2. **Identify all forces and moments**: Include gravitational forces, normal forces, frictional forces, applied forces, and reaction forces at supports or connections.
3. **Draw the forces and moments**: Use arrows to represent forces (pointing in the direction of application) and curved arrows for moments.

Types of Forces and Support Reactions

Forces can be broadly categorized into:

- **Concentrated Forces**: Applied at a single point (e.g., a weight or applied load).
- **Distributed Forces**: Spread over an area or volume (e.g., pressure or uniformly distributed load).

While Support reactions (the response to applied force) depend on the type of support:

- **Pinned Support**: Restricts movement in two directions but allows rotation (provides two reaction forces).
- **Fixed Support**: Restricts all movement and rotation (provides two reaction forces and a moment).
- **Roller Support**: Restricts movement in one direction (provides one reaction force).

Dynamics of Rigid Bodies

Having discussed applied forces we now move on to dynamics, which includes the study of forces and torques under both static and dynamic equilibrium conditions. We will address this topic with a focus on the behavior of objects that do not deform under applied loads.

Key aspects include inertia, which measures the resistance of an object to changes in its state of motion or rest. Rotational kinematics deals with the motion characteristics of objects rotating about an axis, such as angular velocity and angular acceleration. In addition, planar motion analysis helps to understand how objects move within a two-dimensional plane, involving both translational and rotational motion components.

Inertia

Inertia is a basic property of matter that measures an object's resistance to alterations in its state of motion. It is directly related to the mass of the object in translational motion, where more massive objects require more force to change their velocity. This concept is known as rotational inertia or moment of inertia. The moment of inertia depends on the distribution of mass relative to the axis of rotation.

$I = \int r^2 \, dm$ where **r** si the perpendicular distance from the axis of rotation to a small element of mass dm, and **dm** is a small element of mass of the object.

Consider a figure skater who alternates between holding their arms close to their body and extending them outward. When the arms are close to the body, the skater's moment of inertia is lower because the mass is concentrated closer to the axis of rotation, allowing for faster spinning. By extending the arms, the moment of inertia increases due to the greater distance of the mass from the axis, thus slowing the rotation. This example demonstrates how the moment of inertia affects rotational speed, highlighting the importance of mass distribution relative to the axis of rotation.

Rotational Kinematics

Rotational kinematics deals with the motion of bodies that rotate about an axis without considering the forces that cause this motion. Key parameters in rotational kinematics include angular velocity (ω) and angular acceleration (α), which describe the speed of rotation and the rate at which the angular velocity changes over time, respectively. The basic relationships in rotational kinematics mirror those in linear motion but are expressed through angular quantities:

- Angular velocity:
$$\omega = \frac{d\theta}{dt}$$

Here, θ represents the angular displacement in radians, and dt is the change in time. Angular velocity is typically measured in radians per second (rad/s).

- Angular acceleration:
$$\alpha = \frac{d\omega}{dt}$$

This means that angular acceleration is the derivative of angular velocity with respect to time, indicating changes in the speed of rotation. It is measured in radians per second squared (rad/s²).

Planar Motion

Planar motion considers movements that occur within a single plane and is crucial in analyzing systems where objects move in two dimensions. This analysis combines translational and rotational motion principles to provide a complete picture of how forces and torques affect the motion of an object within a plane. The equations of motion for a rigid body undergoing planar motion can be broken down into translational (linear) and rotational components:

- **Translational**: F=ma where F is the net force applied, m is the mass of the body, and a is the acceleration.
- **Rotational**: τ=Iα where τ is the net torque, I is the moment of inertia, and α is the angular acceleration.

Consider a playground seesaw as a simple example to understand planar motion. When children sit on the ends of the seesaw, it undergoes both translational and rotational motion:

Translational Motion: As one child sits down, the seesaw's center of mass shifts, moving vertically under the force of the child's weight.

Rotational Motion: The weight of the child, acting at a distance from the pivot (fulcrum), creates a torque

Moments and Couples

A moment, often referred to as torque, is the rotational effect produced by a force acting at a distance from a point of rotation. The moment M can be calculated using: M=F·d where F is the force and d is the perpendicular distance from the point of rotation to the line of action of the force.

Imagine using a wrench to loosen a bolt. Exert a force at the end of the wrench, while the wrench itself acts at a certain perpendicular distance from the center of the bolt.

A couple is defined by two parallel forces of equal strength but opposite direction, where the lines along which these forces act are not the same. The critical characteristic of a couple is that its effect is purely rotational without causing any translational motion, regardless of the point about which moments are calculated.

The formula to calculate the Couple si the same of moments

Now, consider a situation where two people apply equal and opposite forces at the opposite ends of a horizontal bar that is free to rotate around its center. This creates a **couple**. The bar will rotate around its center without moving laterally because the forces cancel each other out in terms of linear motion but create rotational motion because they are applied at a distance from the rotation point.

Equilibrium of Rigid Bodies

Let's return to the subject of equilibrium for a moment and see how to ensure it for a rigid structure such as a beam. It is necessary for both the total net force and the total net moment (torque) to be zero. Let's see how this principle is used to analyze and identify the support reactions of a beam subjected to a uniform load.

A beam of length L is supported at both ends and bears a uniform load w (force per unit length).

We need to determine the reactions (forces) at the supports to ensure that the beam is in equilibrium.

Set the equilibrium conditions:

1. The sum of the vertical forces $\sum F\_y=0$
2. The sum of the moments about any point (e.g., left support) $\sum MA=0$

Application in Mechanical Systems

Moreover, statics principles guide the design of mechanical systems where components must resist forces without motion. This includes everything from bridges and building supports to stationary machinery components.

Bridge Design Example:

In a simply supported bridge with a central load, you would:

1. Draw the FBD.
2. Apply the equilibrium equations to solve for support reactions.
3. Use these reactions to determine internal forces and moments.

Structural Analysis

In practice, statics is applied in the analysis of structures, from simple beams to complex architectural forms. This involves calculating forces at various points in a structure, ensuring it can withstand the loads it encounters without failure. Engineers use static analysis to determine the forces within trusses, beams, and other structures, ensuring they can safely carry expected loads through methods like free-body diagrams and the method of joints.

Advanced Structural Analysis Techniques:

1. **Method of Sections**: By cutting through the truss and considering the equilibrium of one portion, engineers can solve for the forces acting on individual members. This method is particularly useful for complex trusses where direct methods are cumbersome.

2. **Deflection Analysis**: Techniques such as:
   - the double integration method
   - Macaulay's method numerical
   - finite element method (FEM)

   are used to calculate deflections in beams and frames.

3. **Influence Lines**: These are essential for analyzing structures subjected to moving loads, such as bridges or crane beams. Influence lines enable engineers to determine the maximum effect of a moving load at specific points in a structure

4. **Buckling Analysis**: The Euler formula helps predict the load at which a column will buckle under axial compression, which is directly relevant to ensuring the safety and stability of structures.

5. **Stress Concentration Analysis**: This involves understanding the points within a material or component where stress is higher than the average.

6. **Safety Factor Calculation**: Calculating and applying safety factors appropriately ensures that structures can withstand a reasonable amount of overload without failing.

As You know, all these calculations, help ensure that structures will support the anticipated loads and also maintain acceptable limits of rigidity and comfort.

Dynamics, Kinematics, and Vibrations: Solving Real-World Problems.

Dynamics: Forces and Motion

We have talked about dynamics applied to static bodies; now we will explore the concepts of dynamics applied to forces and motion.

Dynamics focuses on the forces that cause motion and the resulting movements. Newton's Second Law of Motion, which we previously introduced, serves as the cornerstone of dynamics:

$$\mathbf{F}_{net} = m \cdot \mathbf{a}$$

where m is the mass of the object and a is the object's acceleration.

Kinematics: Describing Motion

Both fields of kinematics provide a framework for describing and predicting the behavior of moving objects. You have already seen how rotational kinematics applies to objects rotating around an axis. Linear kinematics applies to objects moving along a path in space. It involves the description of motion without considering the forces that cause it. The primary kinematic variables are position (r), velocity (v), and acceleration (a). These variables are interconnected as follows:

- r: This is a vector that represents the location of an object in space relative to a reference point.
- v: Velocity is defined as the rate of change of position. It is a vector quantity, meaning it has both magnitude and direction. Mathematically, velocity is the first derivative of position with respect to time:

$$\mathbf{v} = \frac{d\mathbf{r}}{dt}$$

- a: Acceleration is the rate of change of velocity. It is also a vector quantity and is given by the first derivative of velocity with respect to time, or the second derivative of position with respect to time:

$$\mathbf{a} = \frac{d\mathbf{v}}{dt}$$

Vibrations: Oscillatory Motion

We may speak of two types of vibration and we'll do it. But first a simple definition:

vibration is the repeated back-and-forth movement of an object around a central point. Imagine a guitar string being plucked. The string moves rapidly back and forth from its resting position, creating a vibration. This movement can affect structures and machinery, sometimes causing wear and tear or even failure if not properly managed.

Free Vibrations: In free vibrations, a system oscillates without any external forces acting on it, following an initial disturbance. The natural frequency (ω) of such a system is a crucial parameter, which depends on the system's stiffness (k) and mass (m). The natural frequency is given by the formula:

$$\omega_n = \sqrt{\frac{k}{m}}$$

Forced Vibrations: In forced vibrations, the system is driven by an external periodic force. The system's behavior under forced vibrations largely depends on the frequency of the applied force relative to its natural frequency. When the external force's frequency aligns with the system's natural frequency, resonance occurs, which can result in large amplitude oscillations if damping is minimal.

Natural Frequencies

Vibrations are nothing more than repeated back-and-forth movements of an object around a central point, as we just mentioned. For these vibrations to propagate, they have a certain frequency at which they oscillate. This frequency can significantly impact the system's performance and stability. What's the point?

Every mechanical system has natural frequencies at which it tends to oscillate if not externally forced or damped. These are inherent properties determined by the system's mass and stiffness distribution. Identifying these frequencies is crucial because resonance can occur if the system is forced to vibrate at its natural frequency, potentially leading to catastrophic failures. Engineers use modal analysis to determine the natural frequencies, ensuring that operational frequencies do not coincide with them.

Here are a few examples:

Bridges: The Tacoma Narrows Bridge collapse in 1940 is one of the most famous examples of resonance leading to structural failure. The bridge oscillated violently in windy conditions due to its design not accounting adequately for aerodynamic forces at its natural frequency. This resulted in a dramatic collapse and emphasized the importance of considering natural frequencies in structural engineering.

Buildings During Earthquakes: Buildings have natural frequencies that can be excited by the frequencies of ground motions during an earthquake. You will must design buildings to withstand seismic forces by ensuring that the building's natural frequencies do not resonate with the predominant frequencies of likely earthquake motions. This is often achieved by using base isolators and other damping systems that alter the natural frequencies of the structure.

Automobiles: In automotive engineering, the natural frequencies of various components (such as the engine, body frame, and suspension system) are critically analyzed. For instance, resonance in the suspension system can lead to excessive oscillations, affecting the car's stability and comfort. Designers ensure that these frequencies do not coincide with the engine's operating frequencies or those generated by the wheels on the road.

Aircraft Wings: Aircraft wings are designed to have high natural frequencies to avoid resonance with the frequencies of the engines and aerodynamic loads experienced during flight.

Machinery and Rotating Equipment: In industrial settings, rotating machinery such as turbines, compressors, and pumps must be designed to operate away from their natural frequencies. If these machines were to operate at or near a natural frequency, they could experience excessive vibrations leading to premature wear, damage, and potentially catastrophic failures.

Damping

Damping refers to the mechanisms through which vibrational energy is dissipated in a system. We just said how it's crucial for controlling vibrations, as it reduces the amplitude of oscillations, enhancing system stability and longevity. In engineering design, different types of damping are considered, such as viscous damping, where the damping force is proportional to the velocity, and structural damping, which involves energy dissipation through material hysteresis.

So, damping dissipates energy and critically affects how a system responds. This is also true for external disturbances. A highly damped system may respond slowly to the external force, thereby reducing the amplitude of the oscillations, while a system with low damping may exhibit large amplitude oscillations, especially if the driving frequency approaches the system's natural frequency.

Real-World Application: Analyzing a Simple Pendulum

Consider a simple pendulum, a classic example that illustrates dynamics, kinematics, and vibrations. A pendulum consists of a mass mmm suspended by a string of length L, which swings due to the force of gravity.

- **Kinematics of the Pendulum**: The position of the pendulum bob can be described by the angular displacement θ(t). The velocity and acceleration are derived from θ(t) and its derivatives.
- **Dynamics of the Pendulum**: Applying Newton's Second Law in the tangential direction gives:

$$mL\frac{d^2\theta}{dt^2} + mg\sin(\theta) = 0$$

For small angles, sin(θ)≈θ, simplifying to:

$$\frac{d^2\theta}{dt^2} + \frac{g}{L}\theta = 0$$

This is a simple harmonic motion equation with a natural frequency:

$$\omega_n = \sqrt{\frac{g}{L}}$$

- **Vibrations**: The solution to the simplified equation is:

$$\theta(t) = \Theta_0 \cos(\omega_n t + \phi)$$

where Θ is initial angular displacement, and φ is the phase constant. This describes the oscillatory motion of the pendulum. This describes the oscillatory motion of the pendulum.

Interactive Section: Real-World Engineering Challenges in Statics and Dynamics

To bridge the gap between theoretical knowledge and practical application, this section introduces a series of interactive problems derived from real-world scenarios in mechanical engineering. These problems are designed to engage and challenge you, enhancing your understanding of statics and dynamics through hands-on learning and application.

Structural Stability Challenge: Bridge Design

Scenario: You are tasked with designing a pedestrian bridge over a busy street. The bridge must be capable of supporting up to 2000 kilograms per square meter during peak hours without yielding.

Interactive Problem: Using the principles of static equilibrium, calculate the necessary dimensions and materials for the bridge's main support beams. Consider factors such as the maximum allowable stress for steel and the overall aesthetic appeal of the structure.

Learning Outcome: This problem helps you apply force equilibrium and moment calculations to ensure the structural integrity and safety of large-scale constructions.

Rotational Dynamics: Wind Turbine Efficiency

Scenario: A company seeks to increase the efficiency of their wind turbines located in a region with variable wind speeds.

Interactive Problem: Analyze the angular velocity and angular acceleration required to optimize power output at different wind speeds. Utilize the concepts of rotational kinematics to suggest modifications to the turbine blade design to maximize efficiency.

Learning Outcome: This exercise demonstrates how dynamic analysis can be utilized to enhance the performance of energy systems in real-time environmental conditions.

Automotive Safety: Crash Impact Analysis

Scenario: As part of a safety engineering team, you need to design a new crumple zone for a car that minimizes the impact force during a frontal collision at 50 km/h.

Interactive Problem: Calculate the force distribution and the required deformation distance for the crumple zone using principles of dynamics and energy conservation. Consider the vehicle's mass, speed, and desired deceleration to minimize injury to passengers.

Learning Outcome: Through this problem, you'll understand the application of dynamics in designing safer vehicles by manipulating force interactions and energy transformations during accidents.

Material Handling: Conveyor System Design

Scenario: Design a conveyor system capable of transporting 500 kilograms of material per minute across a factory floor, involving an inclined section at 30 degrees.

Interactive Problem: Determine the necessary torque and the motor power specifications for the conveyor system, applying the principles of planar motion and static friction.

Learning Outcome: This task will enhance your ability to apply static and dynamic principles to solve for unknowns in systems that require both linear and rotational motion components.

Vibration Control: Building Oscillations

Scenario: A new skyscraper needs a damping system to reduce oscillations caused by earthquakes and wind.

Interactive Problem: Using the concept of natural frequencies and damping, design a damping system that reduces the amplitude of vibrations effectively. Calculate the necessary parameters based on the building's dimensions and expected environmental forces.

Learning Outcome: This problem introduces you to the critical role of dynamics in civil engineering, particularly in the context of natural disaster mitigation and structural design.

Each of these interactive problems aids in applying theoretical principles and also encourages a deep understanding of how these principles impact the design and functionality of mechanical systems in real-world settings. Engaging with these scenarios will sharpen your problem-solving skills and prepare you for the practical challenges faced in the field of mechanical engineering.

Practice

Describe how you would construct a free-body diagram for a simply supported beam with a concentrated load in the center.

Determine the support reactions for a beam that spans 8 meters and carries a uniformly distributed load of 500 N/m.

What is the magnitude and direction of the third force for a system in static equilibrium, if three forces act on a point and two of them are known (3N at 0 degrees and 4N at 90 degrees)?

Explain the significance of Newton's First Law of Motion in the analysis of static equilibrium.

Using the principles of statics, determine the moment around the base of a flagpole caused by a 200 N force acting at a 45-degree angle to the pole's side, 2 meters above the base.

Measure the tension in the cable when a crane lifts a 1500 kg load at a constant speed.

Calculate the bending moment at a point 3 meters from the left support in a 10-meter long bridge when a load of 10,000 N is applied at its center.

Determine the force exerted by the ground on a ladder leaning against a frictionless wall with a person standing halfway up. The ladder weighs 15 kg and is 5 meters long, while the person weighs 70 kg.

Explain the impact on the magnitude of the moment when the distance from the pivot point is doubled for a specific force.

In circular motion, derive how linear velocity relates to angular velocity.

Calculate the angular momentum of a disc with a radius of 0.4 meters and a mass of 2 kg rotating at 12 rad/s.

Explain the consequences of angular momentum conservation using the example of a figure skater spinning and pulling their arms in.

A mass of 5 kg is attached to a spring with a constant of 500 N/m. Calculate the system's natural frequency.

How does adding damping to a system affect its response to external vibrations? Explain with reference to a practical application.

Modeling a car traveling over a bump as a single degree of freedom system, discuss the steady-state and transient responses.

Explain the role of the damping coefficient in determining the behavior of an underdamped system.

Using the concept of rotational kinematics, calculate the final angular velocity of a wheel (initially at rest) that accelerates uniformly at 2 rad/s² over a period of 5 seconds.

Describe the force analysis involved when a vehicle tows a trailer, including considerations of static and kinetic friction.

If a pendulum swings with an initial angle of 30 degrees and a length of 2 meters, determine its speed at the lowest point of the swing.

Evaluate the stability of a vertical column under a compressive load of 1000 N, given its buckling load capacity is 1500 N.

Chapter 9: The Strength of Materials: From Theory to Practical Applications

	Formula
Euler's Formula for Buckling	$P_{cr} = \dfrac{\pi^2 EI}{(KL)^2}$
Slenderness Ratio	$\lambda = \dfrac{K \cdot L}{r}$
Thermal Strain	$\epsilon_{thermal} = \alpha \Delta T$
Thermal Stress	$\sigma_{thermal} = E \alpha \Delta T$
Modified Euler Buckling with Thermal Stress	$P'_{cr} = P_{cr} - \sigma_{thermal} A$
Normal Stress	$\sigma = \dfrac{F}{A}$
Shear Stress	$\tau = \dfrac{F_{shear}}{A}$
Bearing Stress	$\sigma_{bearing} = \dfrac{F}{A_{bearing}}$
Axial Stress	$\sigma_{axial} = \dfrac{F}{A}$
Hooke's Law (Normal Stress-Strain)	$\sigma = E \cdot \epsilon$
Hooke's Law (Shear Stress-Strain)	$\tau = G \cdot \gamma$

Advanced Topics: Buckling, Indeterminate Systems, and Combined Loading

Buckling

If we talk about critical failure modes for structural elements subjected to compressive stresses, Buckling is the protagonist. It occurs when compression leads to a sudden sideways deflection. Engineers must be able to predict the critical load at which a column or beam will buckle to ensure safety and structural integrity.

Have a look at some key concepts:

Euler's Formula for Buckling: Known as a critical in this field, this concept predicts the load at which a slender column will buckle, or suddenly bend under axial compression.

In other words, think of having a thin wooden stick, like a long ruler, in an upright position. If you press down with a small weight on the top of the ruler, it stays straight. But if you keep adding weight, at some point the ruler suddenly bends and collapses to the side. Euler's formula helps predict the exact amount of weight that will cause this phenomenon.

It is given by:

$$P_{cr} = \frac{\pi^2 EI}{(KL)^2}$$

where E is the modulus of elasticity, I is the area moment of inertia, L is the unsupported length of the column, and K is the column effective length factor.

Slenderness Ratio: The slenderness ratio is a measure of a column's susceptibility to buckling. It is defined as the ratio of the column's effective length to its radius of gyration.

Imagine you have two poles of the same height: one is a thin bamboo stick, and the other is a thick wooden post. Even though they are the same height, the thin bamboo stick is more likely to bend and buckle under pressure compared to the thick wooden post. This difference in their tendency to buckle is captured by the slenderness ratio.

Imagine you have two poles of the same height: one is a thin bamboo stick, and the other is a thick wooden post. Even though they are the same height, the thin bamboo stick is more likely to bend and buckle under pressure compared to the thick wooden post. This difference in their tendency to buckle is captured by the slenderness ratio.

Just as we observed with the ruler, which might buckle under too much weight, the slenderness ratio in the example of the bamboo stick and wooden post similarly illustrates how structural dimensions influence susceptibility to buckling. The thin bamboo stick, much like a slender ruler, has a high slenderness ratio, making it more likely to bend and buckle under less pressure because its height is significant relative to its thickness. Conversely, the thick wooden post, akin to a shorter or thicker ruler, exhibits a lower slenderness ratio and is inherently more stable and resistant to buckling.

To misure Slenderness Ratio You can use this formula:

$$\sqrt{\frac{I}{A}}$$

where A is the cross-sectional area. A higher slenderness ratio indicates a higher risk of buckling.

In addition to the impact of slenderness ratio, there are other critical factors that significantly influence the buckling behavior of columns: the conditions at the column ends and the material properties of the column itself.

- **End Conditions**: The conditions at the ends of the column—whether fixed, pinned, or free—significantly influence the effective length and thus the buckling resistance.
- **Material Properties**: Different materials will react differently under compressive stress due to their unique properties like yield strength and Young's modulus

Thermal Stresses in Structural Elements

We have seen that materials can deform under certain loads, but the application of a load is not the only way a material can undergo a change.

Temperature variations can cause materials to expand or contract, which can introduce stresses known as thermal stresses. These stresses can greatly affect the performance and structural integrity of materials, especially when combined with mechanical loads.

Thermal Expansion and Stress

Materials generally expand when heated and shrink when cooled. The extent of this expansion or contraction is determined by the material's coefficient of thermal expansion (CTE), a material property that quantifies how much a material expands per degree of temperature change. The formula to calculate thermal strain is:

$$\epsilon_{thermal} = \alpha \Delta T$$

where α is the coefficient of thermal expansion and ΔT is the change in temperature. The resulting thermal stress, if the deformation is constrained, can be expressed as:

$$\sigma_{thermal} = E \cdot \epsilon_{thermal}$$

where E is the Young's modulus of the material.

Impact on Buckling

Thus, we repeat that for structural elements subject to buckling, such as our columns or beams, thermal stresses can alter the critical buckling loads. The Euler equation for buckling can therefore be modified to also include the effect of thermal stresses

$$P_{cr,thermal} = P_{cr} - \sigma_{thermal} \cdot A$$

where P_{cr} is the critical load calculated without thermal effects, σ is the thermal stress, A represents the cross-sectional area of the column.

In applications such as bridge construction, pipeline design, and electronics manufacturing, where components may experience significant temperature fluctuations, engineers must account for thermal expansion and stresses. Designs typically include allowances for expansion and contraction, such as expansion joints in bridges or flexible mounts in pipelines, to accommodate these changes without inducing undue stress or distortion.

Additionally, in environments where temperature changes are rapid or extreme, materials with low coefficients of thermal expansion, such as certain ceramics or composite materials, may be chosen to minimize the development of thermal stresses.

Indeterminate Systems

Where the simple application of equilibrium conditions is not sufficent to determine the internal forces, these systems are given a different name: they are called indeterminate.

In order for them to be solved, we rely on additional information such as material properties and displacement conditions.

- **Method of Superposition** It: involves separating the effects of different loads and supports to simplify complex structures into simpler, determinate systems. The results from each simplified system are then superposed to find the actual internal forces and reactions. This method assumes linear behavior of the materials, making it applicable only where the deformations are directly proportional to the applied loads.

    Imagine a continuous beam over multiple supports with distributed loads. Let's start by considering each segment of the beam as a separate system. We calculate the reactions and internal forces for each segment and then combine them to get an overall view of the structure.

- **Compatibility Conditions:** In indeterminate structures, simply satisfying equilibrium is not enough; the deformations also need to be compatible with the geometric constraints of the structure. This involves ensuring that the deformations at different parts of the structure do not contradict each other and that the overall geometry of the structure is maintained. For example, in a continuous beam, the deflection at the support points must be the same from both sides of the support to maintain structural integrity.

Guide on Resolving Indeterminate Systems

To thoroughly understand the process of solving indeterminate systems, we must delve deeper into each step, ensuring mastery over the intricate dynamics of structural analysis. Building on the concepts we've introduced, let's enhance our approach by examining the detailed methodologies.

1. Identifying Internal Forces and Constraints:

First, it's imperative to identify all internal forces and constraints within the structure. This involves a meticulous examination of each segment, considering every potential interaction:

- Establish equilibrium equations for both forces and moments for each segment. Recall our discussion on the continuous beam over multiple supports; similarly, treat each segment independently to simplify the analysis.
- Ensuring every force and constraint is accounted for provides a solid foundation for further analysis, reinforcing our control over the structural behavior.

2. Applying the Method of Superposition:

As previously mentioned, the Method of Superposition is crucial for breaking down complex structures into simpler, determinate systems. This technique allows for:

- Separating the structure into its fundamental components, treating each as an independent entity.
- Individually solving for the effects of loads and supports on these components. Remember, as we solve each component, we integrate these results through superposition to synthesize a comprehensive understanding of the entire structure's internal forces and reactions.
- It's important to note that this method presumes linear material behavior, suitable for common materials like steel and wood, where deformations are proportional to applied loads.

3. Establishing Compatibility Conditions:

Beyond simple equilibrium, compatibility conditions are the important ones for indeterminate structures:

- Develop compatibility equations that ensure deformations align with the structural geometry. Referencing our continuous beam example, consider how deflections at support points must align seamlessly to preserve structural integrity.
- This meticulous verification ensures that all parts of the structure cohesively support one another, preventing any contradictions in the overall geometry and maintaining the integrity of the entire system.

4. Utilizing Material Properties:

Do not forget to integrate material properties as well. This will make the results of our compatibility equations even more accurate and reliable:

- Integrate properties such as the modulus of elasticity into your calculations to predict how materials will react under stress.
- Assuming a linear elastic behavior is generally effective for typical construction materials, facilitating straightforward and predictable analysis.

Combined Loading

Structural elements are often subjected to multiple types of loading simultaneously: tension, compression, bending, shear, and torsion. If we want to be realistic, we need to simulate applications that are true to reality, where obviously these forces do not act individually one by one, but obviously simultaneously.

This leads us to adapt sophisticated analysis techniques such as:

- **Stress Transformation:** This involves using mathematical tools to change the reference axes of the stress components at a point to simplify analysis. Mohr's circle, a graphical method, is particularly useful in transforming and visualizing the state of stress at a point under combined loading conditions. By plotting normal and shear stresses, engineers can determine the stresses acting on inclined planes, which is crucial for designing components that must withstand complex stress configurations.
- **Principal Stresses:** are the max and min normal stresses at a point, occurring at specific orientations where the shear stress is zero. Determining these stresses helps in assessing the strength and failure criteria of materials under complex loading. This involves calculating the eigenvalues of the stress tensor, which provide the magnitudes of the principal stresses.

These concepts are foundational for:

- Designing machinery parts that must operate under multiple simultaneous forces.
- Structural health monitoring where different load types impact the longevity and safety of structures.
- Aerospace and automotive industries, where components often face complex stress states due to various operational loads.

Deeper Understanding of Stress, Strain, and Structural Integrity

To summarize:

We began by exploring the phenomenon of buckling, which occurs when excessive compressive force causes a sudden lateral deflection of a component. Additionally, we examined how thermal stresses play a significant role in structural behavior. Temperature variations can cause the expansion or contraction of materials.

Moving on to indeterminate systems, which cannot be solved by static equilibrium alone, we introduced the method of superposition and compatibility conditions. Furthermore, we considered the different types of loads acting on structures - such as tension, compression, bending, shear, and torsion - highlighting how real-world applications often involve complex interactions of these forces.

Building on our comprehensive exploration of structural behavior under various loading conditions and the crucial concepts of buckling and thermal stresses, we now turn our focus to understanding the fundamental nature of stress. Stress, as the internal resistance exerted by a material when subjected to external forces, is a pivotal concept that helps us bridge the gap between theoretical analysis and practical application. This next section will delve deeper into the different types of stress—normal, shear, and bearing—and how they influence material behavior under load. By examining these stress types in more detail, we will gain a clearer understanding of how structures sustain or fail under different forces, setting the stage for advanced discussions on material durability, fatigue, and fracture mechanics.

Stress: Internal Resistance

Stress is described as the internal resistance that a material provides when subjected to external forces. It quantifies the force exerted per unit area within a material and is expressed in pascals (Pa) or megapascals (MPa). Stress can be categorized into three primary types:

- **Normal Stress (σ)**: Occurs perpendicular to the surface and can be tensile (stretching) or compressive (squeezing). The formula for normal stress is: $\sigma = F/A$, where F is the applied force and A is the cross-sectional area.

- **Shear Stress (τ)**: Acts parallel to the surface and is calculated using: $\tau = \frac{F_{shear}}{A}$, where F_{shear} is the shear force. Shear stress results in slippage and displacement in the material parallel to the applied force.

- **Bearing Stress**: Arises when a force is transmitted through an area of contact, such as a bolt in a hole. It is given by: $\sigma_{bearing} = \frac{F}{A_{bearing}}$, where $A_{bearing}$ is the contact area.

Stress and Strain Analysis: Material Behavior Under Load

Analyzing axial, shear, and bearing stresses is crucial in mechanical engineering for designing structures and machinery that are safe and efficient. These concepts help predict and prevent potential failures, ensuring the durability and functionality of engineering components.

Axial Stress: Axial stress arises when a force is applied along the object's length, causing tension or compression. This type of stress is fundamental in understanding how structural elements like beams, columns, and shafts will perform under load. The axial stress (σ) is calculated by the formula:

$$\sigma = \frac{F}{A}$$

F is the force applied and A is the cross-sectional area of the material. For example, in columns, axial stress analysis helps ensure they can support the intended load without buckling.

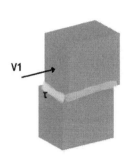

Shear Stress: Shear stress occurs when a force is applied parallel to an object's surface, effectively causing layers or particles to slide against each other. It's crucial for assessing the integrity of fasteners like bolts and rivets, as well as in slip-critical connections in structures. Shear stress (τ) is expressed as:

$$\tau = \frac{V}{A}$$

V represents the shear force and A the area resisting the shear.

Bearing Stress: Bearing stress is a type of stress that occurs when a load is applied over an area in contact, such as in the case of a beam resting on a column. This stress type is particularly relevant in joint design, where components must be sized properly to prevent material deformation or failure. Bearing stress (σ_b) is calculated by:

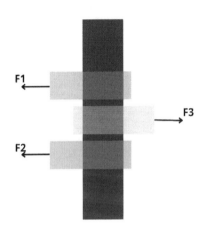

$$\sigma_b = \frac{P}{A}$$

where P is the load and A is the bearing area. Correctly assessing bearing stress is critical for designing machine parts that involve contact and load transmission, such as bearings, joints, and plates.

Understanding the bending moments and shear forces that act upon beams is critical for predicting how these beams will behave under various loads. This involves calculating the stresses and deformations that occur when a beam is subjected to external forces and moments.

Bending moments refer to the response generated in a structural element when an external force is applied, leading to the bending of the element. The shear force is the internal force that acts parallel to the beam's cross-section, resulting from external loads that cause parts of the structure to slide past each other. Deflection, on the other hand, is the displacement of a beam under load.

Fatigue and Fracture Mechanics: Understanding Material Behavior Under Cyclic Loading

Fatigue Analysis

As we know, fatigue occurs when a material is subjected to repeated or fluctuating stresses, typically below the material's yield strength. Over time, these stresses can cause cracks to initiate and propagate, eventually leading to sudden and catastrophic failure. The fatigue life of a material is influenced by its stress amplitude, mean stress, the shape of the component, surface quality, size, and environmental conditions.

Key Concepts in Fatigue

- **Stress-Life (S-N) Curves:** These curves illustrate the relationship between the stress amplitude and the number of cycles failure under cyclic loading. Various materials display different S-N curves, which are utilized to predict a component's fatigue life.
- **Endurance Limit:** Some materials possess a fatigue limit or endurance limit, which is a stress level below which the material can theoretically withstand an infinite number of cycles without failure.
- **Fracture Mechanics:** It deals with the calculation of the stress intensity factor at the tip of a crack and predicts the growth of the crack under various loading conditions. This approach helps in predicting the critical crack size when a crack becomes unstable and growth leads to rapid fracture.

Fatigue and Fracture Prevention

Just to mention, design strategies to mitigate fatigue and fracture include:

- **Improving Material Quality**
- **Surface Treatments**
- **Crack Arrest Features**

Keep them in mind because we'll see again this topic in the next chapter.

Strain: Deformation of Materials

What happens to a material when it passes its load curve but no net break occurs?

We have anticipated this before, strain is the measure that occurs as a result of stress and is dimensionless, representing the relative change in size or shape. This represents a dimensionless measure, quantifying the ratio of dimensional change relative to the original dimension. There are two primary types of strain:

1. **Normal Strain (ϵ):** Occurs due to normal stress and is defined as:

$$\epsilon = \frac{\Delta L}{L_0}$$

where ΔL is the change in length and L_0 is the original length.

2. **Shear Strain (γ):** Arises from shear stress and is measured as the change in angle between two originally perpendicular lines.

Hooke's Law: Stress-Strain Relationship

Hooke's Law explains the relationship between stress and strain in elastic materials, as long as they remain within the proportional limit. It states: $\sigma = E \cdot \epsilon$ where E is the Young's modulus, a material property that measures stiffness. For shear stress and strain, the relationship is: $\tau = G \cdot \gamma$ where G is the shear modulus.

Structural Integrity: Ensuring Safety and Reliability

Structural integrity denotes a structure's capacity to endure its intended load without undergoing failure or significant deformation. Once again main factors influencing structural integrity include:

1. **Material Properties**: Understanding the mechanical properties of materials, such as yield strength, ultimate strength, and toughness, is essential. These properties dictate how materials behave under different loading conditions.
2. **Load Analysis**: Accurately determining the types and magnitudes of loads a structure will encounter is crucial. This includes static loads (constant over time) and dynamic loads (varying with time).
3. **Design Factors**: Applying appropriate design factors, such as safety factors, ensures that structures can tolerate unexpected loads or material defects. Safety factors are typically greater than 1 and account for uncertainties in material properties, load estimations, and potential flaws in construction.

Application Example: Beam Analysis

Think about a beam that is simply supported and bears a uniform distributed load. The stress and strain analysis involves:

1. **Calculating Reactions at Supports**: Using equilibrium equations to find the support reactions.
2. **Shear Force and Bending Moment Diagrams**: Constructing diagrams to visualize how shear force and bending moment vary along the beam.
3. **Stress Analysis**: Using the bending equation: σ=M·cI where M is the bending moment, c is the distance from the neutral axis to the outer fiber, and I is the moment of inertia of the cross-section.
4. **Deflection Calculation**: Ensuring the beam's deflection is within acceptable limits using:

$$\delta = \frac{F \cdot L^3}{48 \cdot E \cdot I}$$

where δ is the deflection, F represents the load, L denotes the length of the beam, E signifies the Young's modulus, and I indicates the moment of inertia.

Practice

Find the critical buckling load of a steel column with a modulus of elasticity of 200 GPa, a radius of gyration of 0.1 meters, and an effective length of 3 meters.

What is the impact of the slenderness ratio on the critical buckling load for a column with one end fixed and the other pinned?

Determine the thermal stress developed in a beam with a coefficient of thermal expansion of $12 \times 10^{-6}/°C$, subjected to a temperature change of 50°C, assuming it is fully constrained.

Analyze a two-span continuous beam using the method of superposition with different load scenarios.

For a combined loading scenario, calculate the principal stresses on a beam subjected to axial compression, bending, and torsion.

Determine the stresses on inclined planes at a point using Mohr's circle for a given state of stress.

Create an experiment to examine the impact of thermal loading on the buckling behavior of columns constructed from various materials.

Discuss the steps to solve an indeterminate truss using the method of sections.

Explain how compatibility conditions are used to solve for displacements in an indeterminate beam under uniform loading.

Evaluate the effects of combined loading on a cylindrical shaft subjected to axial force, bending moment, and torsion.

Propose a design modification to reduce the risk of buckling in a slender column used in a high-temperature environment.

Explain how the choice of material affects the thermal stress and buckling resistance in structural elements.

Calculate the axial stress in a column subject to a 10 kN load with a cross-sectional area of 50 cm².

How would you modify the Euler buckling equation to include the effect of a pre-existing axial stress due to thermal expansion?

Design a structural element that needs to withstand combined bending and torsional loading. Describe the stress analysis you would perform.

Discuss the implications of end conditions on the load-carrying capacity of a pin-ended column under compressive load.

Calculate the thermal expansion for an aluminum bar ($\alpha = 23 \times 10^{-6}/°C$), 1 meter in length, when the temperature increases by 30°C.

Analyze the stability of a column with a variable cross-section along its length under axial compressive load.

For a material with a known yield strength, determine the maximum allowable axial load it can carry without yielding.

Using a case study of a bridge, illustrate how engineers can apply concepts from this chapter to ensure the bridge withstands both static and dynamic loads effectively.

Chapter 10: Advanced Material Properties and Manufacturing Processes

	Formula
Stress Amplitude (S-N Curve)	$\sigma_a = f(N)$
Endurance Limit	$\sigma_{end} = \text{constant}$
Thermal Stress	$\sigma_{thermal} = E\alpha\Delta T$
Stress Intensity Factor (K)	$K = \dfrac{Y\sigma\sqrt{\pi a}}{1}$
Fracture Toughness	$K_{IC} = $ critical value of K
Uniform Corrosion Rate	$R = \dfrac{K \times W}{A \times T \times D}$

In-Depth Look at Material Selection, Properties, and Engineering Processes

Material Fatigue Analysis: Delving into Failure Mechanisms under Cyclic Loading

We've learned that fatigue occurs when materials endure repeated or fluctuating stresses, often below their yield strength. Microscopic cracks may initiate at stress concentrations like notches or sharp corners due to cyclic loading. These cracks can grow over time, leading first to deformation and then to sudden failure. The fatigue life is Impacted stress amplitude, mean stress, the shape and surface quality of the component, its size, environmental conditions, and residual stresses.

Let's delve deeper into the concepts previously discussed.

- **S-N Curves (Stress-Number of Cycles):** S-N curves, or Wöhler curves, are essential in understanding the fatigue properties of materials. By plotting the stress amplitude against the number of cycles to failure, these curves provide a visual representation of how long a material can endure repeated loading before it fails. Typically derived through rigorous testing, the shape of the curve is crucial for engineering applications. As stress amplitude increases, the material's endurance decreases, with the curve steeply descending initially before potentially leveling off if the material has a fatigue limit. This horizontal asymptote suggests that the material does not fail under certain stress levels even after numerous cycles, a property beneficial for designing components subjected to cyclic stresses.

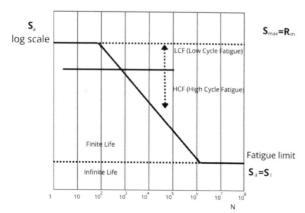

- **Endurance Limit:** unique to certain materials, refers to the stress threshold below which the material can theoretically endure repeated loading indefinitely without suffering from fatigue failure, as previously mentioned. The presence of an endurance limit is particularly significant in engineering applications where components are subjected to frequent or continuous cyclic stresses. Materials such as steel typically exhibit a well-defined endurance limit, making them ideal for critical structural and mechanical components in various industries, including automotive and aerospace. Understanding the endurance limit allows engineers to design safer and more reliable products by ensuring that operational stress levels are kept well below this threshold, thereby extending the service life of the components and minimizing the risk of sudden catastrophic failures.

Having seen how materials behave before on one side of the curve, the side where they never fail, we now look at the phases that characterize when the stess affects them

Fatigue Failure Mechanisms: Fatigue failure mechanisms are critical in understanding how materials behave under repeated stress and are divided into three phases: crack initiation, crack propagation, and final catastrophic failure.

- The initial phase, crack initiation, often begins at sites of stress concentration where minor defects or geometric discontinuities—such as notches, holes, and surface irregularities—exacerbate the local stress environment. This phase is heavily influenced by the material's inherent microstructural characteristics, the quality of the surface finish, and the presence or absence of residual stresses, which may be introduced during manufacturing processes like welding or forming.

  Consider the initial phase of crack initiation as if it were the beginning of a small defect in a mechanical component or the initial carving by waves on a coastal rock. Initially, these imperfections might not appear significant. However, over time, areas under repetitive stress, especially those with small defects or irregular shapes like notches and holes, experience intensified stress. This phenomenon is much like how continuous exposure to water gradually erodes and reshapes rock. The material's microstructure, the quality of its surface finish, and any residual stresses introduced during processes such as welding or forming can exacerbate these stress concentrations, leading to the gradual development and eventual propagation of cracks.

- Once initiated, a crack progresses to the second phase—crack propagation. During this phase, the crack advances incrementally with each load cycle. The rate of propagation is influenced by the applied load amplitude, the material's resistance to crack growth (a property that can be quantified by fracture mechanics), and the operational environment, which may include corrosive or high-temperature conditions that exacerbate the fatigue process.

- The final phase, catastrophic failure, occurs abruptly when the remaining cross-sectional area of the material becomes insufficient to support the applied load, resulting in sudden and often unexpected failure.
  This terminal phase can result in significant damage and, depending on the application, potential safety hazards. The final phase of catastrophic failure can be compared to a cliff finally crumbling after being continuously worn down by the sea. Just as the rock eventually reaches a point where it can no longer withstand the forces exerted by the waves and collapses

Key Concepts in Fracture Mechanics:

- **Stress Intensity Factor (K):** This factor quantifies the stress state near the tip of a crack caused by an external load. You know the third phase? Here, this value tells us exactly how cracking will occur. Stress intensity plays a key role in determining the behavior of a crack under stress, influencing whether it propagates brittlely or ductilely. In brittle fracture, crack propagation is sudden and without significant plastic deformation, while in ductile fracture the material exhibits substantial plastic deformation before failure.
- **Fracture Toughness:** This is a critical property of materials that indicates their ability to resist crack propagation. It represents the critical stress intensity factor (K) value beyond which a material will fail through crack growth. Fracture toughness is a key parameter for material selection and structural design, especially in applications where safety is paramount. Higher fracture toughness means the material can withstand higher stress intensity at the crack tip before it starts to propagate a crack, thus offering greater resistance against fracture.

Strategies for Preventing Fatigue and Fracture:

We've only mentioned them, and now we'll take a closer look. Understanding fatigue and fractures is extremely useful, especially if you also know the practical applications to prevent them.

- **Material Selection and Quality:** Choose high-quality materials with minimal defects to enhance fatigue strength. Superior material quality reduces the likelihood of crack initiation and extends the component's lifespan.
- **Surface Treatments:** Implement techniques like shot peening and surface hardening to induce beneficial compressive residual stresses on component surfaces. These treatments improve fatigue resistance by counteracting tensile stresses that occur during operation.
- **Design Considerations:** Design components with features that reduce stress concentrations, such as fillets and rounded corners. These design choices help distribute loads more evenly and decrease the chances of crack initiation.
- **Crack Arrest Features:** Integrate crack stop holes and other features that can arrest or slow the propagation of cracks. These features enhance durability and safety by preventing sudden failures and extending the component's service life.

Fatigue Life Estimation Methods:

Let's look at the methods used to estimate:

1. **Stress-Life Method (S-N Method):** This traditional method uses the S-N curve to estimate the fatigue life of a material under a given stress amplitude.
2. **Strain-Life Method (ε-N Method):** This approach is used for high-cycle fatigue situations where the strain at critical locations needs to be considered along with the stress. It provides a more comprehensive analysis by incorporating both the elastic and plastic strains in the material.
3. **Fracture Mechanics Approach:** When initial cracks or defects are present, this method predicts the growth of cracks under cyclic loading and estimates the remaining life of the component. It uses concepts such as the stress intensity factor and Paris' law to model crack propagation rates.
4. **Probabilistic Methods:** Given that fatigue can be significantly influenced by variability in material properties, manufacturing processes, and loading conditions, probabilistic methods are employed to account for these uncertainties and provide a statistical probability of failure.

Fatigue analysis is vital in the design and assessment of aircraft components, automotive parts, bridges, and any other structures or components that experience varying loads over their service life. Ensuring that these components can withstand the expected number of cycles without failure is crucial for safety and reliability. Therefore, the following are the criteria when making choices of materials to be used.

Material Selection Criteria

Selecting the right material involves considering various factors:

- **Mechanical Properties:** As previously discussed, properties such as strength, ductility, hardness, and toughness are paramount. These determine how a material behaves under stress and its ability to withstand forces without failing.
- **Physical Properties:** Thermal conductivity, density, and melting point are also crucial, especially in applications involving heat, such as in aerospace and automotive industries.
- **Chemical Properties:** Resistance to corrosion and chemical degradation is vital in environments exposed to harsh chemicals or high moisture, ensuring longevity and reliability.

Engineering Processes

The engineering processes involved in transforming raw materials into engineered products include a wide range of techniques. What is the correlation? Why discuss them now?

Very simple. Each involves specific applications and benefits that are developed precisely through the transformation processes

- **Casting and Molding:** These processes are essential for mass production of metal and plastic components. Die casting is particularly beneficial for metals due to its ability to produce high-volume, complex shapes with excellent surface finish and dimensional consistency. Injection molding for plastics allows for the same complexity and high volume production, utilizing thermoplastic or thermosetting materials which are melted, injected into a mold, and cooled to form the desired shape. These methods are highly efficient and cost-effective for large-scale manufacturing.

Mechanical Properties

Injection-Molded Plastics

- **Tensile Strength:** Varies among materials like polypropylene, polystyrene, and ABS.
- **Flexibility:** High in thermoplastics like polyethylene and polypropylene.
- **Impact Resistance:** Notable in materials like ABS.
- **Thermal Stability:** Certain plastics withstand high temperatures.
- **Chemical Resistance:** Many resist chemicals, crucial for industrial use.
- **Lightweight:** Reduces the weight of the final product.
- **Dimensional Accuracy:** High precision in component dimensions.

Die-Cast Metals

- **Strength:** High, suitable for structural applications.
- **Hardness:** Good wear resistance.
- **Ductility:** Metals like aluminum can be shaped without breaking.
- **Thermal Conductivity:** High, aiding heat dissipation.
- **Corrosion Resistance:** Extends component lifespan.
- **Dimensional Stability:** Excellent precision and consistency.

- **Forming Processes:** Forming techniques such as rolling, forging, and extrusion are employed to shape materials through plastic deformation. Rolling involves passing metal through rollers to reduce its thickness, ideal for creating sheets and plates. Forging uses localized compressive forces to shape metal, often resulting in superior mechanical properties due to refined grain structures and elimination of defects. Extrusion pushes material through a die to produce long objects with a fixed cross-sectional profile, beneficial for producing items with a consistent shape and enhanced strength characteristics.

    Mechanical Properties

    - **Strength:** Rolling enhances the strength of metals by refining the grain structure.
    - **Thickness Reduction:** Allows precise control over the thickness of sheets and plates.
    - **Surface Finish:** Generally results in a smooth surface finish.
    - **Dimensional Accuracy:** Ensures high consistency in thickness and width.

    - **Increased Strength:** Superior mechanical properties due to refined grain structures.
    - **Fatigue Resistance:** Enhanced fatigue strength due to the elimination of internal defects.
    - **Ductility:** Good ductility allows for significant shape changes without cracking.
    - **Toughness:** Improved toughness as a result of a more uniform grain structure.

- **Machining:** This category includes subtractive manufacturing processes that remove material to achieve precise dimensions and finishes. Milling involves the use of rotating cutting tools to

remove material across a variety of axes, providing excellent versatility and precision. Turning, typically performed on a lathe, rotates the workpiece against a cutting tool to shape it to the desired form, often used for cylindrical parts. Drilling creates round holes in material and is indispensable in almost all construction and manufacturing operations. Each machining process is selected based on material characteristics, component geometry, and tolerance requirements.
- Critical for assembling multiple components into functional units, joining processes include welding, soldering, and brazing. Welding involves the fusion of parts through various techniques such as MIG, TIG, and arc welding, suitable for joining metals by melting them together with or without a filler material. Soldering and brazing, however, use a filler metal with a lower melting point than the base materials, ideal for creating joints without altering the properties of the parts being joined. These methods are selected based on the materials' compatibility and the mechanical loads the joint will endure, ensuring structural integrity in the final product

All of these processes are vital in manufacturing, facilitating the creation of reliable and high-quality engineered products efficiently. The choice of process is guided by the material properties, product requirements, and economic considerations to optimize the performance and cost-effectiveness of the final assembly.

Mechanical Properties:

- **Tensile Strength**: The highest stress a material can endure while being stretched or pulled.
- **Yield Strength**: The point at which a material starts to deform plastically under stress.
- **Hardness**: A measure of a material's resistance to deformation or scratching.
- **Ductility**: The ability of a material to deform under tensile stress, often indicated by its capacity to be stretched into a wire.
- **Toughness**: The capacity of a material to absorb energy and undergo plastic deformation without breaking.

Thermal Properties:

Thermal Conductivity: The capacity of a material to transfer heat.

- **Thermal Expansion**: The tendency of a material to expand when heated.

- **Specific Heat Capacity**: The quantity of heat needed to increase the temperature of a unit mass of a material by one degree Celsius.

Electrical Properties:

- **Conductivity**: The capacity of a material to carry electrical current.
- **Resistivity**: The resistance of a material to the flow of electrical current.
- **Dielectric Strength**: The highest electric field a material can endure without breaking down.

Chemical Properties:

- **Corrosion Resistance**: The ability of a material to withstand chemical attack and oxidation.
- **Reactivity**: How readily a material reacts with other substances

Corrosion, Failure Mechanisms, and Their Control

We've discussed how materials deform under load and the importance of measuring fatigue to assess their durability. Now, we'll delve into another form of material degradation that significantly compromises the integrity of components: corrosion. This relentless process can weaken materials over time, potentially reducing their overall strength and effectiveness.

Corrosion Mechanisms and Their Impact

Corrosion represents the chemical or electrochemical reaction between materials, typically metals, and their environments, leading to deterioration and loss of functionality. As we've discussed, the rate and extent of corrosion depend on the material properties and the environmental conditions.

Uniform Corrosion Rate Formula:

$$R = \frac{K \times W}{A \times T \times D}$$

R is the corrosion rate, K is a constant (units vary based on the system, e.g., mm/year for metric), W is the mass loss in grams, A is the area in square meters, T is the time in hours, D is the density of the material in g/cm^3.

- **Uniform Corrosion:** The most common form of corrosion occurs evenly across exposed surfaces. While this type of corrosion is often easy to predict and manage, its implications for structural integrity can be significant, reducing thickness and strength uniformly.
- **Localized Corrosion:** More dangerous than uniform corrosion due to its concentrated nature, localized corrosion includes pitting, crevice corrosion, and stress corrosion cracking. Pitting corrosion, for instance, leads to the formation of small holes in the metal, which can cause severe damage without affecting the overall material significantly, making it hard to detect and predict.

Controlling Corrosion: Protective Measures and Treatments

Preventing and managing corrosion is pivotal for maintaining the structural integrity of materials in harsh environments. Several methods have been developed and refined to control corrosion:

- **Coatings and Linings:** One of the simplest and most effective ways to combat corrosion involves applying protective barriers to the surface of materials. These barriers, such as paints, varnishes, and specially formulated plastic coatings, act as a shield, isolating the material from direct contact with corrosive substances. For example, epoxy coatings are widely used on steel structures for their durability and resistance, and are distinguished by their exceptional properties.
- **Cathodic Protection:** This method is particularly important for protecting large metal structures such as pipelines, ship hulls, and underground storage tanks. Cathodic protection works by attaching a sacrificial anode to the structure, typically made from a metal that corrodes more easily, such as zinc or magnesium. This anode sacrifices itself by corroding in place of the structure it protects, effectively diverting the corrosion process away from the structure's vital components. It is a favored technique for its passive nature and the ability to provide continuous protection without the need for regular maintenance.

Let's make this method more tangible:
Consider your metal structure as a valuable historical monument, and the sacrificial anode as its protective boundary. In environments where corrosion is like weathering—slowly eroding the integrity of the monument—the boundary (anode) absorbs the environmental assault (corrosion), protecting the monument (structure) from decay. Over time, even though the boundary may degrade due to exposure, it can be replaced or restored, ensuring the preservation and long-term resilience of the monument. This setup allows the structure to maintain its integrity and historical significance without suffering irreversible damage.

- **Corrosion Inhibitors:** Corrosion Inhibitors: Widely used in closed systems like cooling towers, boilers, and engine cooling systems, corrosion inhibitors are chemicals added to a fluid to significantly reduce the rate of corrosion. These substances work by forming a thin protective film on the surface of the material or by reacting with potential corrosive agents in the environment to neutralize them. Common inhibitors include phosphates, which are used in water systems to prevent rust, and benzotriazole, used to protect copper and its alloy
- **Material Selection and Design:** Here too, and quite obviously, choosing the right materials and designing structures to minimize corrosion from the outset is a critical strategy. Engineers often select materials and design components based on their environments. For example, stainless steel is preferred for its passive film that regenerates and repairs itself, offering superior resistance in environments prone to corrosion. Similarly, designing structures with fewer crevices and joints can help reduce the areas where corrosive agents can accumulate.
- **Non-Destructive Testing (NDT):** Techniques such as radiography, magnetic particle inspection, and ultrasonic testing are pivotal in the detection of internal and surface defects without damaging the component. Radiography uses penetrating radiation to capture images of a material's internal structure, revealing flaws like cracks or voids. Magnetic particle inspection utilizes magnetic fields to detect surface and near-surface discontinuities in ferromagnetic materials. Ultrasonic testing employs high-frequency sound waves to detect imperfections and measure material thickness, providing critical data without compromising the material's integrity.
- **Material Science Innovations:** The field of material science continually evolves, with research focused on developing new materials that exhibit enhanced properties to resist corrosion and wear. Innovations in superalloys and composites are tailored to perform under extreme conditions. These materials are engineered to offer improved durability and corrosion resistance, vital for applications in harsh environments such as marine, aerospace, and industrial settings. The integration of these advanced materials into structural designs not only enhances longevity but also significantly reduces the lifecycle costs associated with maintenance and repairs.

A Note on Epoxy Coatings

Epoxy coatings are a type of synthetic resin formed by the chemical reaction between an epoxy resin and a hardening agent. When applied to surfaces, they polymerize to form a hard, durable film that adheres strongly to the base material, providing superior protection against corrosion.

You can find them in various industrial and commercial settings. For instance, in the maritime industry, ship hulls are often coated with epoxy to prevent the corrosive effects of seawater and marine organisms. Similarly, in the construction sector, they are used to coat steel bars and reinforcement materials in concrete structures, protecting them from the corrosive effects of moisture and salt, especially in coastal areas.

Why Do They Work So Well?

The effectiveness of epoxy coatings stems from their chemical properties. During the curing process, epoxies form a dense cross-linked network that not only gives them great strength and durability but also makes them highly resistant to chemical attacks and degradation. This barrier is impermeable to water and many corrosive substances, making it an ideal choice for challenging environments.

Additionally, ease of application is another benefit of these coatings; they can be applied as a liquid to new or existing structures using brushes, rollers, or sprayers. Once cured, the coating forms a seamless surface that not only protects but also, as in the automotive industry, provides a high-quality finish.

On the Downside, Not So Negative

These resins are truly a top remedy against corrosion. While they are durable, they are not completely maintenance-free. They may require periodic inspections and possible reapplications, especially in extremely harsh environments.

Integrating epoxy coatings into the broader framework of corrosion control strategies enhances the durability and reliability of materials across a wide range of applications. By understanding and utilizing the unique properties of epoxies, industries can significantly extend the life of their assets and reduce maintenance costs over time.

Practice

Basic Material Fatigue: Calculate the stress amplitude for a steel beam that lasts for 1,000,000 cycles given the S-N curve equation $\sigma a = 1000 \times N^{-0.1}$

Determining Endurance Limit: Given a material with an S-N curve flattening out at a stress level of 250 MPa, determine the endurance limit of the material.

Fatigue Crack Growth Calculation: If a crack in an aluminum component grows at a rate given by $\frac{da}{dN} = 10^{-12} \Delta K^4$ where ΔK is in MPam $\sqrt{m}$, calculate the number of cycles for a crack to grow from 1 mm to 2 mm under a stress intensity range of 15 MPam$\sqrt{m}$.

Stress Intensity Factor Problem: For a plate with a through-thickness crack of 30 mm subjected to a stress of 100 MPa, calculate the stress intensity factor using the formula K=$\sigma\sqrt{\pi a}$.

Fracture Toughness Design Question: If a material has a fracture toughness of 50 MPam$\sqrt{m}$, determine the maximum allowable crack size when the operating stress is 120 MPa.

Comparative Analysis of Materials: Discuss how you would choose between two materials with different S-N curves for a component subjected to cyclic loading of 500 MPa stress amplitude.

Material Selection for Thermal Stress Resistance: Calculate the thermal stress in a material with a Young's modulus of 210 GPa and a coefficient of thermal expansion of $12 \times 10^{-6}/°C$ when subjected to a temperature change of 150°C.

Handling Corrosion in Design: Propose a protective measure for a metal structure operating in a marine environment and justify your choice based on corrosion types.

Failure Analysis Scenario: Given a scenario where a component failed after 10,000 cycles at a stress of 600 MPa, use the S-N curve method to analyze the likely reasons for failure and suggest design improvements.

Fatigue Life Estimation Exercise: Using the stress-life method, estimate the fatigue life of a component subjected to a stress amplitude of 350 MPa using the given S-N curve equation.

Thermal Expansion Impact Analysis: Calculate the change in length of an aluminum bar (original length = 2 m, coefficient of thermal expansion = $23 \times 10^{-6}/C$), when the temperature is increased by 100°C.

Design for Variable Loading: A beam is subjected to bending, torsion, and axial loading simultaneously. Outline the steps you would take to analyze the combined stresses using stress transformation techniques.

Critical Load Calculation for Buckling: Calculate the critical load for a steel column ($E = 200\ GPa, I = 500 \times 10^4\ mm^4$, effective length = 2 m) using Euler's formula.

Understanding End Conditions in Buckling: Discuss how different end conditions affect the buckling load of a column and propose a design change to enhance stability.

Calculating Slenderness Ratio: For a column with a radius of gyration of 50 mm and an effective length of 2 m, calculate the slenderness ratio and assess the risk of buckling.

Manufacturing Process Selection: Compare and contrast the effects of die casting and injection molding on the mechanical properties of a product.

Corrosion Rate Calculation: If a metal plate loses 0.5 grams of mass due to corrosion over an area of 0.1 m^2 in 1000 hours, calculate the uniform corrosion rate using the given formula.

Stress Concentration Factor Analysis: Explain how you would modify a design to reduce stress concentrations around a hole in a tensile load-bearing plate.

Fatigue and Fracture Prevention Techniques: Suggest and justify a surface treatment process for an aircraft wing component to enhance its fatigue life.

Practical Application of Fracture Mechanics: Given a crack length of 20 mm in a critical structural component and a stress intensity factor calculation, discuss the steps you would take to assess the need for immediate replacement or repair.

Chapter 11: Fluid Mechanics for the Practical Engineer

	Formula
Continuity Equation	$A_1 V_1 = A_2 V_2$
Bernoulli's Equation	$p + \frac{1}{2}\rho v^2 + \rho g h = \text{constant}$
Navier-Stokes Equations	$\rho \left(\frac{\partial \mathbf{v}}{\partial t} + \mathbf{v} \cdot \nabla \mathbf{v} \right) = -\nabla p + \mu \nabla^2 \mathbf{v} + \mathbf{f}$
Reynolds Number (Re)	$Re = \frac{\rho v D}{\mu}$
Froude Number (Fr)	$Fr = \frac{v^2}{gD}$
Mach Number (Ma)	$Ma = \frac{v}{a}$
Hydraulic Grade Line (HGL)	$z + \frac{p}{\gamma}$
Energy Grade Line (EGL)	$z + \frac{p}{\gamma} + \frac{v^2}{2g}$

Essential Principles and Problem Solving in Fluid Dynamics

Fundamental Concepts

Fluid dynamics is the branch of physics that studies the behavior of liquids and gases in motion. It focuses on understanding how fluids flow in various conditions and under different forces, analyzing factors such as velocity, pressure, density, and temperature changes.

The study of fluid mechanics is rooted in several key principles that are essential for the FE Mechanical Exam:

1. **Continuity Equation**:

$$A_1 V_1 = A_2 V_2$$

This formula asserts that, under conditions of steady, incompressible flow, the mass flow rate remains unchanged across different pipe cross-sections. Here, 'A' represents the cross-sectional area and 'V' indicates the velocity of the fluid. This principle is vital for analyzing pipe systems and fluid transport.

2. **Bernoulli's Equation**:

   Bernoulli's equation links the pressure, velocity, and elevation of a fluid in a streamline to a constant, providing a powerful tool for examining fluid flow in terms of energy conservation. It's particularly useful in applications involving fluid flow through ducts or over wings.

   $$p + \frac{1}{2}\rho v^2 + \rho g h = \text{constant}$$

   Where p is the liquid pressure, ρ represents the density, v indicates the velocity, g is the acceleration due to gravity, and h is the height above a reference point.

3. **Navier-Stokes Equations**: These equations describe the motion of viscous fluid substances. They are derived from Newton's second law and provide a comprehensive description of fluid flow. In three dimensions, they are expressed as:

   $$\rho\left(\frac{\partial \mathbf{v}}{\partial t} + (\mathbf{v} \cdot \nabla)\mathbf{v}\right) = -\nabla P + \mu \nabla^2 \mathbf{v} + \mathbf{f}$$

   where ρ is the fluid density, v is the fluid velocity vector, $\frac{\partial v}{\partial t}$ is the partial derivative of the velocity with respect to time, (v·∇)v represents the convective acceleration, $-\nabla P$ is the gradient of the pressure, μ represents the fluid's dynamic viscosity, $\nabla^2 v$ is the Laplacian of the velocity field, indicating viscous diffusion, f is the body force per unit volume acting on the fluid.

After exploring the fundamental principles and equations that govern fluid dynamics, we can now transition to how these theoretical concepts are applied in real-world scenarios using advanced technologies.

- **Computational Fluid Dynamics (CFD)**: The Navier-Stokes equations are fundamental in CFD simulations, which enable engineers to model and analyze the complex behaviors of fluid flows. CFD employs numerical methods and algorithms to solve and analyze problems involving fluid flows. Without computers, the calculations needed to simulate the interaction of liquids and gases could not be performed. The role of Navier-Stokes in this context is to provide the detailed, physics-based modeling of fluid momentum, which includes components for velocity, pressure, density, and viscosity at any point in the fluid.

  In practical CFD applications, the Navier-Stokes equations help predict weather patterns, design aircraft and automotive components, optimize HVAC systems, and develop medical applications like blood flow through arteries. They allow engineers to visualize flow patterns, test new ideas in a virtual environment, and make modifications before any physical prototypes are built.

- **Simplified Models for FE Mechanical Exam:**

For the purposes of the FE Mechanical Exam, the full complexity of the Navier-Stokes equations often exceeds the scope of what is tested. Instead, simplified models of these equations are more relevant. These simplified versions focus on specific scenarios that are common in engineering problems, such as:

Flow in a Straight Pipe: This involves assuming a steady, incompressible flow that simplifies the Navier-Stokes equations to a more manageable form where the flow velocity and pressure gradients can be more easily calculated.

Flow Around Simple Shapes: For example, the flow around spheres or cylinders can often be approximated under certain conditions to provide insights into drag forces and flow separation without solving the full three-dimensional Navier-Stokes equations.

Dimensionless Numbers and Their Significance in Fluid Dynamics

As previously discussed, fluid dynamics relies heavily on certain fundamental principles, among which dimensionless numbers. They play a crucial role in analyzing fluid behavior under various conditions. The Reynolds, Froude, and Mach numbers are particularly significant for predicting flow patterns, transitions between flow regimes, and the effects of compressibility. Here we expand upon these concepts to deepen our understanding:

1. **Reynolds Number (Re):**

The Reynolds number (Re) is the ratio of inertial forces to viscous forces in a fluid flow, and it is expressed as

$$Re = \frac{\rho v D}{\mu}$$

Where ρ denotes the fluid density, v represents the flow velocity, D is the characteristic length (like the diameter of a pipe), and μ signifies the dynamic viscosity of the fluid.

This dimensionless number indicates whether the flow will be laminar or turbulent. Laminar flow happens at low Reynolds numbers, where viscous forces are dominant, resulting in a smooth and orderly flow. Conversely, at high Reynolds numbers, inertial forces prevail, leading to turbulent flow characterized by chaotic fluid particle movements.

Understanding the Reynolds number is crucial for engineers as it governs the shift from laminar to turbulent flow, which is essential in various applications:

- **Pipe Flow:** To determine the flow regime within pipelines, engineers utilize the Reynolds number. Specifically, a Reynolds number under 2000 signifies laminar flow, noted for its smooth and orderly nature. Conversely, when the Re surpass 4000, the flow transitions to turbulent, known for its chaotic patterns.
- **Aerodynamics:** In the design of aircraft wings, the Reynolds number helps in determining the airflow characteristics over the wing. Lower Reynolds numbers might indicate laminar flow, which reduces drag but might also reduce lift at certain angles.

2. **Froude Number (Fr):**

The Froude number (Fr) is a dimensionless quantity that compares flow inertia to gravitational forces, defined as

$$\text{Fr} = \frac{v}{\sqrt{gD}}$$

Where v is the flow velocity, g is the acceleration due to gravity, D is the characteristic length (such as the depth of water in an open channel).

This number is particularly significant in contexts where gravity plays a major role in fluid flow.

Understanding the Froude number is essential in various engineering applications:

- **Ship Design:** The Froude number helps predict the wave resistance of ships in water. By scaling ship models accurately to their full-size counterparts, designers can forecast performance at sea, particularly regarding wave creation and resistance. This is crucial for ensuring efficient and effective ship designs.
- **Open Channel Flow:** In designing canals, spillways, and other open channels, the Froude number indicates whether the flow is subcritical (Fr < 1) or supercritical (Fr > 1). Subcritical flows are slower and more controlled, while supercritical flows are faster and resemble rapids. This distinction is vital for designing hydraulic structures that manage flooding or optimize water conveyance.

3. **Mach Number (Ma)**

The Ma is another dimensionless figure representing the ratio of an object's velocity to the speed of sound in the fluid around it. It is calculated as

$$\text{Ma} = \frac{v}{a}$$

, where v is the object's speed and a is the speed of sound in that medium.

The Mach number is particularly significant in scenarios involving compressible flow, such as:

- **Aircraft Speed:** In aerospace engineering, the Mach number is crucial for designing aircraft to operate at different flight regimes. Subsonic aircraft (Ma < 1) and supersonic aircraft (Ma > 1) require different design approaches due to the effects of shock waves and variations in aerodynamic forces. Transonic (Ma ≈ 1) and hypersonic (Ma > 5) flights introduce additional complexities that must be addressed to ensure stability and performance.
- **Rocket Propulsion:** For rockets, understanding the Mach number is vital for predicting and managing the effects of shock waves during high-speed travel through the atmosphere.

The Mach number's importance lies in its ability to describe the compressibility effects of the fluid. Different Mach regimes—subsonic, transonic, supersonic, and hypersonic—have distinct characteristics that impact the design and performance of aircraft and spacecraft.

Boundary Layer Theory

In fluid mechanics, the boundary layer is the thin layer of fluid next to a surface where viscous effects are significant. As fluid moves over a solid surface, friction between the fluid and the surface causes the fluid's velocity to decrease from the free stream to zero at the surface. This velocity gradient defines the boundary layer, which can be categorized into laminar and turbulent layers depending on the flow conditions. In other words, imagine you're painting a wall with a brush. As you move the brush along the wall, the paint right at the bristles (touching the wall) moves the slowest because it's sticking to the wall. As you move slightly away from the wall, the paint moves a bit faster, and farther away from the bristles, it moves even faster still. This variation in paint speed from the wall outwards is like what happens in the boundary layer of fluid flow.

There are two type of layer:

1. **Laminar Boundary Layer**:

   - **Characteristics**: Fluid particles flow in smooth, parallel layers with minimal mixing between adjacent layers. This type of boundary layer exhibits predictable flow patterns and relatively low energy dissipation.
   - **Flow Conditions**: Laminar flow occurs at lower velocities and is more stable. It is typically present when the Reynolds number (Re), which is the ratio of inertial forces to viscous forces, is below a critical threshold.
   - **Transition**: As the flow velocity increases or other conditions change (such as surface roughness or disturbance), the laminar boundary layer could transition to a turbulent one.

2. **Turbulent Boundary Layer**:

   - **Characteristics**: Involves chaotic mixing of fluid particles across various scales, leading to increased momentum transfer and higher energy dissipation compared to laminar flow.
   - **Flow Conditions**: Turbulent boundary layers are common in high-velocity flows or when the Reynolds number exceeds a critical value.
   - **Impact**: The chaotic nature of turbulent flow enhances mixing and heat transfer rates but also increases aerodynamic drag and energy loss.

Why are they important?

This knowledge about boundary layers facilitates smarter and more efficient design and operation of systems in various fields of engineering, highlighting the importance of fluid dynamics in modern technological advances. We see in detail that:

- **Aerodynamic Drag**: Mastering boundary layers is vital for reducing aerodynamic drag on vehicles. Proper management of these layers leads to enhanced fuel efficiency and improved overall performance by smoothing the flow of air around the vehicle's surfaces.

- **Heat Transfer**: Boundary layers play a pivotal role in heat transfer processes. A thorough understanding of how these layers behave is essential for effective thermal management in various engineering applications, from HVAC systems to aerospace engineering.

- **Engineering Design**: Analyzing how boundary layers transition and behave under different conditions allows engineers to refine and optimize designs. This knowledge is crucial for developing more efficient fluid flow systems, including pipelines and aircraft surfaces, ensuring they operate optimally under various flow conditions.

Hydraulic and Energy Gradient Lines

HGL

The Hydraulic Grade Line represents the level to which water will rise in piezometers due to the hydrostatic pressure within the system. It is a visual depiction of the pressure head (or piezometric head) at various points along a pipeline or channel. The HGL is defined mathematically as:

$$\text{HGL} = z + \frac{p}{\gamma}$$

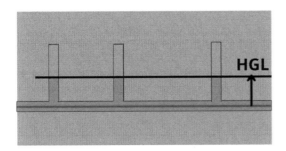

where z is the elevation head, p is the pressure at the point of interest, and γ is the specific weight of the fluid. In practical terms, the HGL indicates the energy available due to pressure in the fluid and is used to assess the potential for fluid flow from one point to another under gravity.

EGL

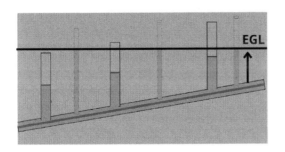

The Energy Grade Line, on the other hand, represents the total mechanical energy of the fluid per unit weight along the flow path. It includes the velocity head, pressure head, and elevation head. The EGL is expressed as:

$$\text{EGL} = z + \frac{p}{\gamma} + \frac{v^2}{2g}$$

Significance in System Design

The distance between the EGL and HGL at any point in the system is equal to the velocity head $\frac{v^2}{2g}$ and represents the kinetic energy per unit weight. This separation is indicative of the fluid's velocity: a larger distance suggests higher velocity. In designing efficient piping systems and open channels, engineers must ensure that the EGL does not drop below certain critical points to avoid issues such as cavitation or excessive pressure loss.

This difference is significant for several reasons:

1. **Velocity Indication**: As we have learned, a larger gap between the EGL and HGL suggests a higher velocity. This is crucial for engineers to understand the flow characteristics within the system. Higher velocities, while potentially beneficial for certain applications like power generation (where rapid flow through turbines generates more power), can also pose risks such as erosion of pipes or channels, increased noise levels, and greater potential for damaging vibrations.

2. **Energy Efficiency**: Efficient energy management within a fluid system hinges on maintaining an optimal velocity that balances dynamic fluid behavior with system integrity and function. This is essential in minimizing energy losses due to friction and other resistive forces, as previously discussed in our examination of Bernoulli's principle and the conservation of energy in fluid streams.

3. **Avoiding Cavitation and Pressure Anomalies**: Engineers must design systems to ensure that the EGL remains appropriately above critical points along the system's length. If the EGL drops excessively, particularly below the HGL, it can lead to undesirable phenomena such as cavitation. Cavitation occurs when the local fluid pressure drops below the vapor pressure, causing vapor bubbles to form, which can cause significant mechanical damage upon their collapse.
4. **System Safety and Maintenance**: Understanding and controlling the position of the EGL relative to the HGL helps in predicting and managing pressure fluctuations and ensuring the structural safety of the system. This understanding helps in the placement of pumps, valves, and other control devices to regulate flows effectively and safely, especially in systems operating under high pressures or varying elevation profiles.

Practical Applications

- **Piping Systems**: In complex piping networks, such as those in municipal water supply or industrial applications, maintaining an appropriate EGL is essential to prevent pump failure and to design efficient and safe systems.
- **Open Channels**: For waterways and irrigation channels, the HGL and EGL help in designing channels that minimize energy losses due to friction and other disturbances, ensuring optimal flow characteristics.

Fluid Properties and Characteristics:

Density (ρ): crucial parameter in fluid mechanics and it defined as mass per unit volume. It is expressed as:

$$\rho = \frac{m}{V}$$

where m is mass and V is volume. The density of a fluid affects its buoyancy and pressure distribution.

Viscosity (μ): Viscosity measures a fluid's resistance to flow. It's the most important to determining the fluid's internal friction. Fluids with high viscosity, like honey, flow slowly, whereas fluids with low viscosity, such as water, flow effortlessly. Dynamic viscosity (μ) and kinematic viscosity (ν) are related by:

$$\nu = \frac{\mu}{\rho}$$

Surface Tension: This property is significant in phenomena such as capillarity and droplet formation. It results from cohesive forces between fluid molecules at the surface.

Energy, Impulse, and Flow Considerations

Understanding the foundational properties such as density, viscosity, and surface tension of fluids lays the groundwork for delving deeper into how these properties influence and interact with energy transfer and impulse-momentum dynamics in fluid systems. With these properties defining how fluids behave under various conditions, we are now prepared to explore more complex interactions in fluid mechanics. This includes how energy is converted and utilized in systems involving pumps, turbines, and fans, and how the principles of impulse and momentum are critical in analyzing fluid motion.

- **Energy Transfer in Fluid Systems**: Often examined through pumps, turbines, and fans, where mechanical energy is converted from and to fluid energy, influencing how systems are designed and analyzed.

$$F\Delta t = \Delta(m\mathbf{v})$$

where F is the applied force, Δt denotes the change in time, m is the mass of the object, v is the velocity of the object, and Δ(mv) is the change in momentum of the object.

While exploring the interplay of energy, impulse, and flow dynamics in fluid mechanics, it's crucial to understand how these concepts manifest in practical scenarios, painting a clear and vivid picture of their applications. In fluid mechanics, energy transfer is a core concept, particularly visible in devices like pumps, turbines, and fans.

- Pumps: Imagine a pump as a heart within a system, propelling fluid through pipes by increasing the fluid's energy. The pump adds energy to the fluid, increasing both its pressure and kinetic energy, which helps overcome resistance due to pipe friction and elevations in the system.
- Turbines: Conversely, turbines act like reverse pumps. Picture water rushing from a high-level reservoir through a turbine. Here, the fluid's potential energy, due to its elevation, and kinetic energy from its motion, are converted into mechanical energy as the water spins the turbine blades. This mechanical energy is then often converted into electrical energy in power plants.
- Fans: Fans work similarly by accelerating air, a fluid, effectively transferring electrical energy (from an electrical motor) into kinetic energy (the motion of the air). In a room, a fan circulates air, enhancing airflow and distributing thermal energy more evenly.

Problem-Solving Strategies

Problem-solving in fluid dynamics often involves applying theoretical concepts to real-world scenarios. Here are some strategies that can be helpful:

- **Dimensional Analysis**: Dimensional analysis is a powerful mathematical method used to simplify complex physical phenomena into dimensionless parameters. This technique helps engineers and scientists to understand the relationships between different physical quantities without getting bogged down by units. For instance, by using dimensional analysis, one can determine how changes in scale or flow velocity affect the performance of hydraulic systems without constructing detailed prototypes.

    Consider the design of a water slide in an amusement park. Dimensional analysis can help predict how different water flow rates will affect rider speed and safety across various slide designs. By forming dimensionless groups like the Reynolds number, which relates inertial and viscous forces, engineers can scale up their findings from small models to the actual structures.

- **Similitude and Modeling**: Similitude is a concept in fluid dynamics used to ensure that physical models or simulations accurately replicate the conditions that will occur in the actual application. This involves adhering to geometric, kinematic, and dynamic similarities. Models, whether physical or computational, are scaled versions of full-scale applications, and understanding similitude principles is key to accurate prediction.

Using these principles, engineers can create scaled models of ships in water tanks to study hydrodynamic performance without the need to first build a full-sized ship. By applying the correct scaling laws based on the Reynolds and Froude numbers, the observations made on the model can be accurately extrapolated to predict the behavior of the actual ship at sea.

Energy, Impulse, and Flow: Tools for Mechanical Engineering

Fluid Mechanics for the Practical Engineer

In fluid dynamics, energy exists in several forms: potential energy, kinetic energy, and internal energy. The fundamental principle governing energy in fluid systems is the conservation of energy, often applied through the Bernoulli equation and the First Law of Thermodynamics.

1. Potential Energy (PE): Potential energy in a fluid system is due to its elevation above a reference level. It is given by: PE=ρgh,

where ρ denotes the density, g represents the acceleration due to gravity, and h is the height above the reference point.

2. Kinetic Energy (KE): Kinetic energy is associated with the fluid's velocity and is expressed as:

$$KE = \frac{1}{2}\rho v^2$$

where v is the fluid velocity.

3. Internal Energy (U): Internal energy represents the energy due to the molecular structure and temperature of the fluid. It is significant in thermodynamic analyses but often less emphasized in simple fluid dynamics problems.

Bernoulli's Equation: Energy Conservation in Fluid Flow

That's represent a statement of the conservation of mechanical energy in a flowing fluid. For an incompressible, non-viscous fluid, Bernoulli's equation is expressed as:

$$P + \frac{1}{2}\rho v^2 + \rho gh = \text{constant}$$

where P is the pressure, ρ is the density, v is the velocity, and h is the height above a reference point.

This equation indicates that the sum of pressure energy, kinetic energy, and potential energy remains constant along a streamline. Bernoulli's equation is pivotal in understanding how energy is transferred within fluid systems and is extensively used to analyze fluid behavior in various engineering applications.

First Law of Thermodynamics: Energy Balance

The principle of energy conservation, known as the First Law of Thermodynamics, asserts that energy cannot be created or destroyed, but only transformed from one form to another. In fluid systems, this law can be applied to control volumes to analyze energy transfer. The energy balance equation for a control volume is: $\Delta E_{in} - \Delta E_{out} = \Delta E_{system}$ where ΔE_{in} is the energy input to the system, ΔE_{out} is the energy output from the system, ΔE_{system} is the change in the system's energy.

Practical Applications: Pump and Turbine Efficiency

Energy analysis is critical in designing and optimizing fluid machinery like pumps and turbines. As we mentioned previously these devices either add energy to the fluid (pumps) or extract energy from the fluid (turbines).

1. **Pump Efficiency**: A pump adds energy to a fluid, increasing its pressure and velocity. The efficiency (η) of a pump is defined as the ratio of the hydraulic power delivered to the fluid ($P_{hydraulic}$) to the mechanical power input to the pump.

$$\eta = \frac{P_{hydraulic}}{P_{mechanical}}$$

2. Hydraulic power can be calculated using:

$$P_{hydraulic} = \rho g Q H$$

where Q is the volumetric flow rate and H is the head added by the pump.

3. **Turbine Efficiency**: As mentioned, a turbine harnesses energy from a fluid by transforming kinetic and potential energy into mechanical work. A turbine's efficiency and hydraulic power have the same calculations as the pump.

Optimizing Energy Efficiency

We know how Maximizing efficiency is vital in this field so here's a few key strategies:

- **Minimizing Friction Losses**: Reducing friction losses in pipes and channels by using smoother materials and optimizing flow rates.
- **Enhancing Pump and Turbine Design**: Utilizing advanced design techniques to improve the efficiency of pumps and turbines.
- **Implementing Energy Recovery Systems**: Incorporating systems like regenerative braking or energy recovery ventilators to reclaim wasted energy.

Impulse and Fluid Dynamics

Only mentioned it before the impulse-momentum equation provides critical insights into how forces interact with fluids over time. It equates the change in momentum of a fluid to the impulse applied, which is particularly useful in situations involving sudden changes in fluid velocity, such as water hammer in pipelines or thrust generation in jet engines. By understanding this relationship, engineers can predict and control the effects of dynamic forces on fluid systems.

The impulse-momentum theorem states that the change in momentum of a fluid is equal to the impulse applied to it. This relationship is given by: I=Δp where I is the impulse and Δp is the change in momentum. For a fluid, this can be expressed as:

$$\int_{t_1}^{t_2} \mathbf{F}\, dt = m \Delta \mathbf{v}$$

F represents the applied force, m is the mass of the fluid, and Δv denotes the change in velocity.

For a control volume with steady flow, the impulse-momentum equation can be written as:

$$\sum \mathbf{F} = \dot{m}(\mathbf{v}_{out} - \mathbf{v}_{in})$$

where m˙ is the mass flow rate, v_{out} is the velocity at the outlet, and v_{in} is the velocity at the inlet.

Practical Applications: Nozzles and Diffusers

Building on our understanding of the impulse-momentum equation and its critical role in analyzing fluid dynamics, we now apply this fundamental principle to specific engineering devices such as nozzles and diffusers. By examining how forces interact with fluids over time, particularly during sudden changes in velocity, we gain deeper insights into the performance of these devices.

1. Nozzles: A nozzle is designed to accelerate a fluid by converting pressure energy into kinetic energy. Understanding the impulse-momentum relationship is essential for analyzing the performance of nozzles, particularly in jet engines and rocket propulsion.

- **Flow through a Nozzle**: As fluid enters a nozzle, its velocity increases while its pressure decreases. The impulse-momentum equation can be used to calculate the force exerted by the fluid on the nozzle. The flow direction through a nozzle is given by:

$$F_x = \dot{m}(v_{\text{out}} - v_{\text{in}})$$

    where F_x represents the axial force, m˙ is the mass flow rate, and v denotes velocities at the inlet and outlet.

- **Example Calculation**: Consider a nozzle where water enters at 10 m/s and exits at 30 m/s, with a mass flow rate of 5 kg/s. The force exerted on the nozzle can be calculated as:

$$F_x = 5\,\text{kg/s} \cdot (30\,\text{m/s} - 10\,\text{m/s}) = 100\,\text{N}$$

2. Diffusers: A diffuser is used to decelerate a fluid, converting kinetic energy into pressure energy. This is critical in various engineering applications, such as HVAC systems and turbines.

- **Flow through a Diffuser**: As fluid enters a diffuser, its velocity decreases while its pressure increases. The force exerted by the fluid on the diffuser is analyzed in a manner similar to nozzles but with reversed velocity changes.

- **Example Calculation**: Suppose air enters a diffuser at 50 m/s and exits at 20 m/s, with a mass flow rate of 3 kg/s. The force exerted on the diffuser can be calculated as:

$$F_x = 3\,\text{kg/s} \cdot (20\,\text{m/s} - 50\,\text{m/s}) = -90\,\text{N}$$

The negative sign signifies that the force direction is opposite to the flow direction.

Practice

Continuity Equation Application: What is the ratio of the initial to the final cross-sectional area of a pipe if the water velocity increases from 3 to 6 m/s?

Bernoulli's Principle Problem: Determine the pressure difference between two locations in a horizontal pipe with water flow. At Point A, the velocity is 2 m/s and the pressure is 120 kPa, while at Point B, the velocity is 4 m/s. Assume the fluid is incompressible and has a density of 1000 kg/m³.

Navier-Stokes Simplification: For a steady, incompressible flow in a straight pipe, derive the simplified form of the Navier-Stokes equations assuming no body forces.

CFD Simulation Scenario: Describe how you would set up a CFD simulation to analyze the airflow over a car's wing at high speed. Include considerations for boundary conditions and mesh refinement.

Reynolds Number Calculation: Compute the Reynolds number for water traveling in a pipe with a diameter of 0.3 m and a velocity of 2 m/s at 25°C. Assume the dynamic viscosity of water is 0.9×10^{-3} Pa·s.

Laminar to Turbulent Transition: At what velocity does a pipe flow of water at 20°C transition from laminar to turbulent if the pipe diameter is 0.1 m?

Froude Number Application: A ship model, 6 meters in length, is tested in a towing tank at a speed of 2 m/s, with gravity at 9.81 m/s². Determine the Froude number and explain its implications for the full-scale ship design.

Mach Number Significance: Explain the importance of the Mach number in the design of supersonic jets and provide an example of how shock waves affect jet performance.

Boundary Layer Discussion: Discuss the effects of a laminar versus a turbulent boundary layer on heat transfer and skin friction for an aircraft wing.

Energy Grade Line Interpretation: For a hydraulic system, if the EGL and HGL are 3 meters apart at a certain section of the pipe, calculate the fluid velocity at this section.

Hydraulic Grade Line in Practice: Given a pipeline with varying diameters, how would changes in the HGL help in identifying possible sites for energy recovery or pump placement?

Stress-Life Curve Usage: Explain how an S-N curve can be used to predict the fatigue life of a bridge under cyclic loads.

Endurance Limit for Material Selection: For a material with a known endurance limit, determine the maximum allowable stress amplitude for a part expected to experience 10 million cycles.

Stress Intensity Factor Problem: Calculate the stress intensity factor for a crack in an airplane's fuselage subjected to a stress of 30 MPa. Assume the crack length is 2 cm.

Fracture Toughness Decision Making: How would knowledge of a material's fracture toughness influence the choice of materials for high-stress components in aerospace?

Fatigue Failure Case Study: Analyze a failure in a car suspension spring subjected to fluctuating stresses, using the S-N curve for the material.

Stress-Life Method Example: If a metal component is subjected to a stress amplitude of 150 MPa, and the corresponding S-N curve data shows a fatigue life of 500,000 cycles, explain the potential long-term implications for a vehicle's operation.

Dimensional Analysis in Fluids: Utilize dimensional analysis to derive a relationship between pressure drop, fluid velocity, pipe diameter, and fluid viscosity in a pipe flow.

Similitude in Model Testing: Discuss how the concept of similitude is used in hydraulic modeling, providing an example with a scale model of a dam.

CFD in Automotive Design: Describe a scenario where CFD would be critical in optimizing the cooling system of a high-performance car engine. Include considerations for mesh size and turbulence modeling.

Chapter 12: Thermodynamics and Heat Transfer: Core of Mechanical Engineering

	Formula
First Law of Thermodynamics	$\Delta U = Q - W$
Isothermal Process (Ideal Gas)	$PV = nRT$
Adiabatic Process (Ideal Gas)	$PV^\gamma = \text{constant}$
Isobaric Process	$Q = \Delta U + P\Delta V$
Isochoric Process	$\Delta U = Q$ (no work done)
Carnot Efficiency	$\eta = 1 - \frac{T_{cold}}{T_{hot}}$
Fourier's Law of Heat Conduction	$q = -kA\frac{\Delta T}{\Delta x}$
Newton's Law of Cooling (Convection)	$q = hA(T_s - T_\infty)$
Stefan-Boltzmann Law (Radiation)	$E = \sigma A T^4$
Combined Heat Transfer	$q_{total} = q_{cond} + q_{conv} + q_{rad}$

Fundamentals and Applications of Thermodynamics

Basic Concepts

The study of energy is the central focus of thermodynamics, work, and heat, and how they interact within physical systems. Key concepts include systems, properties, states, processes, and cycles.

1. **System and Surroundings**:
    - A system refers to the specific portion of the universe we are focused on analyzing, whereas the surroundings encompass everything external to the system. Systems can be

open (exchange of mass and energy), closed (exchange of energy but not mass), or isolated (no exchange of mass or energy).
- Open Systems: These systems can exchange both mass and energy with their surroundings. Examples include turbines and pumps, where fluid enters and exits, carrying energy.
- Closed Systems: These systems allow energy transfer but not mass transfer across their boundaries. An example is a piston-cylinder assembly, where the gas inside can exchange heat with its surroundings but no mass enters or leaves.
- Isolated Systems: are those that do not exchange mass or energy with their surroundings. An example is an insulated thermos flask, designed to minimize energy transfer.

2. **Properties**:
   - **Thermodynamic properties** describe the state of a system. are classified into two categories: intensive and extensive.
     - **Intensive properties**, such as temperature and pressure, do not depend on the system's mass and remain constant regardless of how large or small the system is. Other intensive properties include boiling point, melting point, density and specific heat capacity.
     - **extensive properties**, in contrast, depend on the size or amount of matter present in the system. These include volume, total energy, entropy and amount of substance. For instance, the volume of a gas is directly proportional to the amount of gas present: doubling the amount of gas also doubles the volume, if the pressure and temperature remain constant.

   The **state** of a system is described by its properties at a given moment. When all properties are specified, the system's state is fully defined. **State functions** are properties that depend only on the current state, not the path taken to reach that state. This point in important, remember this.

3. **State and Equilibrium**:
   - The state of a system is defined by its properties at a given moment. These include temperature, pressure, volume, and composition. The state of a system can be represented on diagrams, such as pressure-volume (P-V) or temperature-entropy (T-S) diagrams, which help visualize the relationships between different properties.
     - Thermodynamic Equilibrium: A system is in thermodynamic equilibrium when it satisfies three types of equilibrium: thermal, mechanical, and chemical.
   - Thermal Equilibrium: This happens when there is no temperature gradient either within the system or between the system and its surroundings. In thermal equilibrium, there is no net heat flow.
   - Mechanical Equilibrium: A system is in mechanical equilibrium when there are no unbalanced forces acting within it or between it and its surroundings. This means there is no change in pressure across the system, and no parts of the system are moving relative to each other.
   - Chemical Equilibrium: This equilibrium is reached when the system's chemical composition remains constant over time. The rates of any chemical reactions within the system are balanced, with the forward and reverse reactions occurring at the same rate, leading to no net change..

When a system is in equilibrium, we can assume its properties are uniform and unchanging, making it easier to apply thermodynamic principles and equations to predict system behavior.

4. **Processes**:
   - **A process** refers to the transition of a system from one state to another. This transformation includes alterations in the system's properties, like temperature, pressure, and volume. Understanding these processes is essential for analyzing and designing thermodynamic systems. Here are some common thermodynamic processes, each with unique characteristics and applications:
     - **Isothermal Process**: it occurs at a constant temperature. During this process, the system exchanges heat with its surroundings to ensure the temperature remains unchanged.

       They are common in ideal gas calculations and could be represented on a P-V diagram as a hyperbolic curve. They are used in processes like slow compression or expansion of gases where heat transfer is sufficient to maintain constant temperature. The isothermal expansion of an ideal gas in a piston-cylinder assembly, where the gas expands slowly enough to allow heat transfer to keep the temperature constant.
     - **Adiabatic Process**: It occurs without any heat transfer between the system and its surroundings. Any change in the system's internal energy results solely from work done on or by the system.
       Application: Adiabatic processes are critical in understanding rapid processes where there is no time for heat exchange, such as in shock waves and sound propagation. The rapid compression of gas in an engine cylinder, where the process happens too quickly for heat exchange with the surroundings.
     - **Isobaric Process:** It occurs at a constant pressure. The heat added or removed from the system changes its internal energy and volume.
       They are often used in heating and cooling applications where pressure is maintained constant, such as in boilers and condensers. An isobaric process is represented by a horizontal line on a P-V diagram. Heating water in an open container at atmospheric pressure, where the water expands as it is heated but the pressure remains constant.
     - **Isochoric Process:** It occurs at a constant volume. Since the volume remains constant, no work is performed by or on the system, and any heat added or removed alters the system's internal energyThey are used in situations where volume remains constant, such as in rigid containers. On a P-V diagram, an isochoric process is represented by a vertical line. Heating a gas in a sealed, rigid container causes the pressure to increase as the temperature rises, while the volume stays constant.
5. **Cycles**:
   - **A thermodynamic cycle** consists of a series of processes that return a system to its initial state. During a cycle, the system undergoes various changes in its properties, such as temperature, pressure, and volume, but at the end of the cycle, all properties return to their original values.

     Here's a more detailed look at the importance and types of thermodynamic cycles:

Purpose of Thermodynamic Cycles

- **Energy Conversion:** Thermodynamic cycles are used to convert energy from one form to another. For instance, in a heat engine, thermal energy is transformed into mechanical work, whereas in a refrigerator, mechanical work is used to move heat from a cold area to a warm area.
- **Efficiency Analysis:** Cycles assist in evaluating the efficiency of energy conversion processes. By analyzing the work output and heat input of a cycle, we can calculate the efficiency and identify areas for improvement.
- **Repeatability:** Since a cycle returns to its initial state, it can be repeated indefinitely, making it useful for continuous operations in industrial applications.

Types of Thermodynamic Cycles

- **Power Cycles:** These cycles are designed to produce work.
    - **Carnot Cycle:** An idealized cycle that yields the highest possible efficiency for a heat engine operating between two temperature reservoirs consists of two isothermal processes and two adiabatic processes.
    - **Rankine Cycle:** Commonly used in power plants, this cycle converts heat into mechanical work, typically using water and steam as the working fluid. It comprises four processes: isentropic compression, constant-pressure heat addition, isentropic expansion, and constant-pressure heat rejection.
    - **Otto Cycle:** The idealized cycle for spark-ignition internal combustion engines, like gasoline engines, includes two isentropic processes and two constant volume processes.
    - **Diesel Cycle:** The idealized cycle for compression-ignition engines, such as diesel engines. It includes two isentropic processes, one constant-pressure process, and one constant-volume process.
- **Refrigeration Cycles:** These cycles aim to transfer heat from a low-temperature reservoir to a high-temperature reservoir.

Vapor-Compression Cycle: Commonly used in refrigerators and air conditioners, this cycle includes the compression of a refrigerant, heat exchange, expansion, and evaporation to absorb heat from the refrigerated space and release it to the surroundings.

- **Absorption Refrigeration Cycle:** Uses a refrigerant and an absorbent to transfer heat. Unlike vapor-compression systems, it uses heat energy instead of mechanical work to drive the cycle, making it useful for applications where waste heat is available.

Laws of Thermodynamics

We have all heard of the laws of thermodynamics. Maybe some people have heard one or two. The most common idea is that there are three, but there are actually Four principles on which the behavior of thermodynamic systems are governed:

Zeroth Law: The Zeroth Law of Thermodynamics states that if two systems are in thermal equilibrium with a third system, they are in thermal equilibrium with each other. This law establishes the concept of temperature.

First Law of Thermodynamics: The first law, often referred to as the law of energy conservation, states that energy within a closed system is constant. Energy can neither be created nor destroyed; it can only be transformed from one form to another or transferred between systems. Mathematically, it is expressed as:

$$\Delta U = Q - W$$

where ΔU is the change in internal energy of the system, Q is the heat added to the system, and W is the work done by the system on its surroundings.

Second Law of Thermodynamics: This law introduces the concept of entropy, a measure of system disorder or randomness. The second law states that in an isolated system, entropy can never decrease over time. This law highlights the directionality of processes and the inevitable increase in entropy leading to energy dissipation and inefficiency. For practical engineering, this law underscores why no process involving energy conversion is ever 100% efficient.

Third Law of Thermodynamics: The third law states that as the temperature approaches absolute zero, the entropy of a system approaches a minimum, constant value. While this scenario is theoretical, as absolute zero is unattainable, the law provides the foundation for understanding low-temperature behavior in materials and the quantum nature of particles.

Let's clear up some confusion and simplify the understanding of the four fundamental laws of thermodynamics, illustrating them with practical examples

0

Think of it like a social gathering: if person A feels comfortable in the temperature of room X and person B feels comfortable in room Y, and room X and room Y are both adjusted to the same temperature, then A and B would feel comfortable with each other in terms of temperature when they meet. In engineering, this law is essential for the design of thermometers and temperature control systems. For instance, ensuring that components in a computer or any electronic device operate within a specific temperature range to maintain performance and prevent overheating.

1

Now consider a steam turbine: the chemical energy from burning coal is converted into heat energy in the boiler, then into mechanical energy in the turbine, and finally into electrical energy via the generator. This principle is foundational in all energy conversion devices including engines, refrigerators, and power plants, ensuring energy efficiency and optimization in system designs.

2

Entropy is a measure of disorder or randomness, and this law suggests that systems naturally progress toward more disordered states. In practical terms, consider the process of charging and using a battery in a device such as a smartphone or laptop. Once charged, the battery stores electrical energy that is then converted into various forms of energy to power the device's screen, processor, and other components.

When a battery is charged, electrical energy is converted into stored chemical potential energy. However, when the battery is used, not all of this stored chemical energy is converted back into electrical energy. Some of it turns into thermal energy (heat) due to the resistance in the battery and the device's circuits. This heat is not useful for powering the device and represents a loss of usable energy. Thermal energy is more disordered or random than electrical or chemical energy, contributing to a higher entropy state. This is why no energy conversion process is 100% efficient. A portion of the energy input will invariably be converted into forms that are not useful for the intended purpose.

3

The Third Law of Thermodynamics states that as the temperature of a system approaches absolute zero, its entropy—or the measure of disorder within the system—approaches a minimal constant value. To better understand this, let's consider a practical example involving superconductors, which are materials that can conduct electricity without resistance at very low temperatures.

In an MRI (Magnetic Resonance Imaging) machine, superconducting materials are used to create a strong and stable magnetic field necessary for imaging. Superconductors are ideal for this application because they can carry large currents without losing energy as heat, a property that becomes perfect at temperatures near absolute zero.

Imagine a scenario where we want to use a copper wire in an MRI machine instead of a superconductor. At room temperature, copper has some resistance to electrical flow, which means it heats up as electricity passes through it. This heating is due to the internal disorder within the copper's atomic structure, where electrons collide with atoms, losing energy with each collision.

Now, replace the copper with a superconducting material and cool it down close to absolute zero. As the temperature drops, the material's entropy decreases because its atomic particles start to align more orderly, reducing the random movements. At a certain low temperature, the superconductor reaches a state where it has minimal entropy and can conduct electricity without any resistance, thus without any heat loss. This state of minimal entropy and resistance is what superconductors in MRI machines exploit to maintain their efficiency.

Applications of Thermodynamics

Thermodynamics finds applications across a wide range of engineering disciplines. Here, we focus on some of the primary applications relevant to mechanical engineering.

1. **Heat Engines**:
    - Heat engines convert heat energy into mechanical work. Examples include internal combustion engines and steam turbines. The efficiency of a heat engine is expressed as:

    $$\eta = \frac{W_{out}}{Q_{in}}$$

    where W_{out} is the work output and Q_{in} is the heat input.

2. **Refrigeration and Heat Pumps**:
    - Refrigerators and heat pumps transfer heat from a cooler space to a warmer one, using work input. The performance of these systems is evaluated by the coefficient of performance (COP):

    $$\text{COP}_{heat\ pump} = \frac{Q_{out}}{W_{input}}$$

    Where Q_{out} represent the heat delivered to the heated space, and W_{imput} is the work input to the heat pump.

3. **Thermodynamic Cycles**:
    - Thermodynamic cycles such as the Carnot, Rankine, and Brayton cycles are fundamental to power generation and refrigeration. Each cycle has unique characteristics and efficiency, determined by the specific processes involved.

4. **Phase Change**:
    - Understanding phase change is crucial for applications involving refrigeration, heating, and materials science. The energy required for phase changes, such as melting or boiling, is characterized by latent heat.

Practical Example: Analyzing a Heat Engine

Consider an ideal Carnot engine operating between a high-temperature reservoir at 500 K and a low-temperature reservoir at 300 K. The efficiency (η\etaη) of this engine can be calculated using the Carnot efficiency formula:

$$\eta = 1 - \frac{T_{low}}{T_{high}}$$

$$\eta = 1 - \frac{300}{500} = 0.4$$

Substituting the given temperatures:

Thus, the Carnot engine operates at 40% efficiency, highlighting the importance of temperature differences in determining engine performance.

Practical Guide to Heat Transfer

Heat transfer is a fundamental concept in mechanical engineering that involves with the movement of energy due to temperature differences. It occurs in three primary modes: conduction, convection, and radiation. Each mode involves different mechanisms and mathematical descriptions, but they often work together in real-world applications. Recognizing the role of each mode and their interactions is crucial for designing efficient thermal systems.

Conduction

Conduction is the process by which heat energy is transferred through a material without the material itself moving. This mode of heat transfer is typically found in solids, where heat flows from a region of higher temperature to one where the temperature is lower.

Fundamental Law of Conduction: The rate at which heat transfers through a material is dictated by Fourier's law of heat conduction, which can be expressed as follows:

$$q = -k \cdot A \cdot \frac{\Delta T}{\Delta x}$$

where q is the heat transfer rate, k is the thermal conductivity of the material, A is the cross-sectional area, ΔT is the temperature difference, and Δx is the thickness of the material.

Convection

Convection involves the transfer of heat by the physical movement of a fluid (liquid or gas). This process is one of the most important in many engineering applications, including heating and cooling systems, atmospheric dynamics, and oceanic currents. Convection can be categorized into two types:

- **Natural Convection**: Occurs when fluid motion is driven by buoyancy forces arising from density variations due to temperature differences within the fluid.
- **Forced Convection**: Involves fluid being moved by external sources, such as pumps or fans, in a controlled manner.

Mathematical Description: The convective heat transfer can be described by Newton's law of cooling:

$$q = h \cdot A \cdot (T_s - T_\infty)$$

where h represents the convective heat transfer coefficient, A is the area through which heat is transferred, T_s is the surface temperature, and T_∞ is the temperature of the fluid far from the surface.

Convection is exploited in designing radiators, heat exchangers, and even in meteorology for weather forecasting models.

Radiation

Radiation is the transfer of heat by electromagnetic waves and does not require any medium. It is the fundamental method of heat transfer in space and plays a significant role in solar heating and cooling technologies.

- **Basics of Radiation**: The energy emitted by a body due to radiation is proportional to the fourth power of the absolute temperature, as described the Stefan-Boltzmann law:

$$E = \sigma \cdot A \cdot T^4$$

where E is the energy radiated per unit time, σ is the Stefan-Boltzmann constant*, A is the area, and T is the absolute temperature of the body.

*$5.67 \times 10^{-8} \, W/m^2K^4$

Radiation principles are critical in understanding the heat loss in building materials, the design of spacecraft, and the operation of solar panels.

Integrating Heat Transfer Modes

In most real-world engineering problems, all three modes of heat transfer interact. For instance, in a typical household radiator, heat is conducted through the metal body, carried away by convection currents in the air, and also radiated to the surroundings.

Combined Heat Transfer Analysis

Engineers must often calculate the total heat transfer involving all three mechanisms, which requires a comprehensive understanding of how they interact under different conditions.

1. **Heat Transfer Coefficients**:

- To integrate the three modes, engineers use combined heat transfer coefficients that account for conduction, convection, and radiation. These coefficients simplify the analysis by providing a single value that represents the total heat transfer rate.

2. **Energy Balance Equations**:
    - We previously discussed Fourier's law for conduction and Newton's law of cooling for convection. For combined analysis, these equations are used together with the Stefan-Boltzmann law for radiation. An energy balance equation might look like:
    $$q_{total} = q_{conduction} + q_{convection} + q_{radiation}$$
    This combined equation helps in solving complex thermal systems where all three modes are significant.

3. **Thermal Resistance Network**:
    - Engineers often use a thermal resistance network to simplify the analysis. Each type of heat transfer is defined by its thermal resistance, and the combined thermal resistance is the total of these separate resistances. Consequently, the overall heat transfer rate can be calculated using:
    $$q_{total} = \frac{\Delta T}{R_{total}}$$
    R_{total} is the total thermal resistance.

4. **Practical Applications**:
    - **Electronic Cooling**: In electronic devices, heat generated by components is conducted through the device materials, removed by convective airflow, and radiated away. Efficient thermal management requires optimizing all three heat transfer modes.
    - **Building Insulation**: Effective building insulation systems consider conduction through walls, convection due to air leakage, and radiation from surfaces. Engineers design materials and structures to minimize overall heat transfer, enhancing energy efficiency.
    - **Automotive Systems**: In car engines, heat transfer involves conduction from the engine block, convection through coolant fluids, and radiation from engine surfaces. Understanding and optimizing these interactions ensures better thermal control and engine performance.

So we see how all the concepts addressed are fundamental and more importantly are used every day often in things we take for granted.

Practice

Define an Open System: Describe an open system and provide two examples where this type of system applies in mechanical engineering.

Isothermal vs. Adiabatic Processes: Compare and contrast isothermal and adiabatic processes within the context of a piston-cylinder assembly. How does each process affect the internal energy of the system?

First Law of Thermodynamics Problem: A system that is closed has an input of heat totaling 500 kJ and exerts work of 200 kJ. Calculate the alteration in the internal energy.

Thermal Equilibrium Concept: Explain what it means for two systems to be in thermal equilibrium and discuss the implications for heat transfer between them.

Mechanical Equilibrium Application: Consider a container divided by a movable, frictionless piston with gas on both sides at different pressures. Describe how mechanical equilibrium is reached.

Chemical Equilibrium in Engineering: Provide an example of a mechanical system where chemical equilibrium is crucial. What happens if the equilibrium is disturbed?

Isochoric Process Calculation: A gas in a fixed-volume container of 0.5 m³ receives an addition of 150 kJ of heat, which results in a temperature increase. Calculate how the pressure changes from its starting point of 100 kPa.

Rankine Cycle Efficiency Problem: Calculate the efficiency of a Rankine cycle where steam enters the turbine at 3 MPa and 350°C and exits the condenser at 50 kPa. Assume ideal conditions.

Carnot Cycle Question: What is the highest efficiency achievable by a Carnot engine that functions between two thermal reservoirs at temperatures of 500 K and 300 K?

Conduction Heat Transfer: Calculate the heat transfer rate through a wall with a thermal conductivity of 0.8 W/m·K, area of 10 m², thickness of 0.2 m, and temperature difference across the wall of 30°C.

Natural vs. Forced Convection: Discuss the difference between natural and forced convection and provide practical examples where each is used in mechanical engineering systems.

Radiation Heat Transfer Problem: A small object at 600 K emits radiation into surroundings at 300 K. Assuming the emissivity is 0.9 and the surface area is 0.1 m², calculate the net heat loss due to radiation.

Combined Heat Transfer Scenario: An electronic device includes a chip that dissipates 50 W of power and is cooled using a combination of conduction, convection, and radiation. Sketch the thermal resistance network for this system.

Phase Change Calculation: Assuming the latent heat of fusion for ice is 334 kJ/kg, calculate the energy needed to melt 10 kg of ice at 0°C.

Thermal Resistance in Insulation: Calculate the overall thermal resistance for a composite wall comprising three layers with thermal resistances of 0.5, 0.3, and 0.2 $m^2 \cdot K/W$, respectively.

Heat Pump COP Calculation: A heat pump delivers 2.5 kW of heat while consuming 1 kW of electrical power. Calculate its coefficient of performance.

Isothermal Compression in Refrigeration How is the entropy of the system affected during the isothermal compression of a refrigerant, where the pressure rises from 1 bar to 10 bar?

Heat Engine Work Output Problem: Calculate the work output of a heat engine that takes in 500 kJ of heat from a hot reservoir and expels 300 kJ of heat to a cold reservoir.

Properties of Refrigerants: Discuss why certain properties of refrigerants, such as boiling point and specific heat, are important for their selection in refrigeration cycles.

Thermodynamic Property Diagram: Create and label a P-V diagram representing an ideal gas going through an isochoric process and then an isobaric process. Describe the internal energy variations for each part of the process.

Chapter 13: The Future of Mechanical Engineering- Emerging Technologies and Advanced Materials

This chapter focuses on methodologies that are revolutionizing mechanical engineering and reshaping the market as we know it. We are keenly aware that ours is a world in constant evolution, where every strategy employed, every analysis conducted, and every concept applied is under continuous review. This is the very essence of innovation and, upon deeper reflection, the driving force behind scientific revolutions.

To draw a parallel with a previously discussed concept, there exists a balance until a force is applied, instigating a reaction. What, then, is the reaction in this context? It is the revolution: the adoption and discovery of new technologies and methodologies. These are processes that supplement their predecessors and gradually replace them to enhance current performances, ensuring a seamless and motivating transition toward advanced engineering practices.

Revolutionizing Production: Additive Manufacturing

It is coommonly known as 3D printing, has fundamentally transformed how products are designed, developed, and delivered. This technology allows for the layer-by-layer creation of objects from digital models, opening up new realms of possibility in mechanical engineering.

It stands out because it builds objects by adding material, one layer at a time, which is a total contrast to traditional subtractive manufacturing methods that remove material to shape an object. This approach minimizes waste and allows for the creation of complex, customized shapes that would be impossible or impractical with traditional methods.

The process begins with a digital 3D model, which is sliced by software into hundreds or thousands of horizontal layers. These layers are then printed one at a time, fused together typically through the melting of plastic, metal, or other materials by a laser or extruder, depending on the specific technology used (such as Fused Deposition Modeling (FDM), Stereolithography (SLA), or Selective Laser Sintering (SLS)).

One of the most significant advantages of additive manufacturing is its ability to produce complex geometries with a high degree of customization without additional cost as just mentioned. This capability is particularly beneficial in mechanical engineering, where tailored parts can provide improved performance and integration.

Moreover, additive manufacturing excels in rapid prototyping, providing engineers with the ability to quickly turn concepts into functional prototypes. This speeds up the iterative design process, allowing for faster refinement and testing of ideas.

Without considering one of the most interesting and revolutionary parts of this process: parts can be printed in one piece, and this reduces the need for multiple components and assemblies.

This small aspect a) simplifies the production process and b) improves the strength and integrity of the product

Let's see where this techonlogy is employed in the most complex fields:
- In the aerospace sector

For example, a leading aerospace company utilized 3D printing to create a fuel nozzle that was not only lighter than its traditionally manufactured counterpart but also more durable and better performing due to its optimized design for fuel flow.

- In the medical field

Engineers have used 3D printing to create customized implants for patients. These implants are designed to perfectly match the patient's anatomy, improving the comfort and effectiveness of medical treatments. For instance, a patient received a titanium pelvic bone implant that was specifically designed to fit their body, showcasing the potential of additive manufacturing in personalized medicin

Revolutionizing Industries

The future of additive manufacturing in mechanical engineering is vibrant and expansive. Innovations in materials science, such as the development of new metallic alloys and composite materials, are poised to further enhance the capabilities of 3D printers. Engineers and scientists are collaborating to develop materials that are not only more versatile and durable but also capable of conducting electricity, withstanding extreme temperatures, and offering increased flexibility.

Polymer-based materials, traditionally used in 3D printing, are being enhanced to increase their utility. For instance, recent advancements have led to the creation of polymers that are more heat-resistant and capable of maintaining structural integrity under mechanical stress. As previously mentioned, these can be used as components in automobiles. Imagine the day when we can print an entire car, starting with the engine and all its components, and finishing with the bodywork.

In the realm of materials, titanium and aluminum alloys are commonly used for their strength-to-weight ratio, which is crucial in the aerospace and automotive sectors. However, recent developments include stainless steel and nickel superalloys that offer improved properties, such as greater strength and corrosion resistance at high temperatures.

Or consider that by embedding carbon or glass fibers within a polymer matrix, these composites can achieve unprecedented strength and stiffness, paving new paths for construction and industrial applications. Here, we are talking about producing materials for building; one day, we may be able to print materials that we will use to construct buildings and houses.

Or consider the medical field, where 3D printing is making significant strides with the development of biocompatible materials. Think about the stringent standards these materials must meet in terms of regulatory safety and compatibility. Recent innovations include the introduction of resorbable polymers that can be used to create temporary implants in surgery. These polymers gradually dissolve in the body, eliminating the need for a second surgical procedure to remove the implant.

The environmental impact of production is a growing concern that 3D printing can uniquely address. Researchers are focusing on the development of sustainable materials derived from biological sources. For instance, PLA (polylactic acid), made from corn starch, is a biodegradable plastic that has been widely adopted for its low environmental impact. And it is not the only one. We know the impact that renewables have today, and the coverage this field is managing to achieve is incredible and in line with all the standards we have previously discussed. Those working with this technology are committed to development and comprehensive inclusion, respecting the surrounding environment and with a vision far ahead of current standards.

Lastly, moving beyond the boundaries of engineering and venturing into the culinary field, it's fascinating to discover that significant strides are being made in experimenting with innovative techniques for 3D food printing. Among these, one of the most revolutionary involves the possibility of printing meat. This not only opens new frontiers for the food industry but also promises to revolutionize how we think about the production and consumption of meat products. Using living cells cultured in laboratories, this technology allows the creation of tissues that mimic real meat in texture, flavor, and nutritional value, without the environmental and ethical costs associated with traditional meat production.

Automation and Robotics in Manufacturing

In this ever-evolving world of modern manufacturing, automation and robotics have emerged as fundamental forces driving significant enhancements in efficiency, precision, and productivity. These technologies are not just enhancing existing manufacturing methods but are completely transforming the industry, ushering in a new era of innovation and expanded capabilities. Building on the advancements discussed in the previous section on additive manufacturing, automation and robotics further extend the boundaries of what is possible in mechanical engineering, seamlessly integrating with new manufacturing techniques to create smarter, more efficient production workflows. Thus, we see further change and consequently evolution. Initially, we used robots for simple, repetitive tasks. Today, the same robots are employed in different environments, undertaking more complex tasks and facing more challenging problems. Consider this as a step forward and in the direction where tomorrow the robots we have built will be giving directives to those that will build themselves, allowing us to completely step away from these forms of labor. To clarify with an example, consider Amazon's warehouses, where there is virtually no human labor. The robots move and manage logistics through algorithms that control their movements and behavior.

Understanding Automation and Robotics

Automation in manufacturing involves the use of control systems, such as computers or robots, and technology to manage different processes and machineries in an establishment to replace human intervention. But what actually is Robotic?

Basically that is a branch of engineering that involves the conception, design, manufacture, and operation of robots. Tools and machines are often designed to handle repetitive tasks, but modern robotics integrate advanced sensors and software to perform more complex tasks with high precision.

Key Benefits and Applications

1. **Enhanced Productivity:** One of the most significant benefits of integrating robotics and automation is the substantial increase in production rates. Robots can operate 24/7 without fatigue, which not only speeds up the manufacturing process but also delivers higher throughput.

2. **Improved Quality and Consistency:** Automation technologies provide superior precision and repeatability, which means products are manufactured with consistent quality and fewer errors compared to manual production. This consistency is crucial for industries where high precision is necessary, such as in the production of medical devices or aerospace components. Also consider that while robots can perform tasks with higher precision and without fatigue, the displacement of human workers presents a moral dilemma. It challenges engineers and industry leaders to consider how technology should be integrated in a manner that supports human workers rather than replacing them entirely. Fundamentally, the primary role of robots is to assist humans with their tasks, and it is a welcome development, especially when it involves saving physical effort. What raises a moral dilemma, however, is the misinterpretation by some who believe that machines will replace us. In reality, if we look back, many jobs that existed before, such as those on assembly lines, have replaced human labor. This trend will continue; it's inevitable. Yet, once again, this is often viewed from the wrong perspective. In fact, the integration of machines has given rise to new sectors: these machines need to be maintained, programmed, and managed. Hence, for every job position that closes, three new ones open.

3. **Increased Safety:** Automating dangerous or repetitive tasks lowers the risk of injury to human workers, enhancing workplace safety. Robots can perform high-risk activities in hazardous environments, which decreases the likelihood of accidents and improves overall safety. On the other hand, this integration requires rigorous ethical scrutiny to ensure that reliance on robotic systems does not lead to new forms of workplace risk or complacency in safety standards. This dual aspect of robotics—enhancing safety yet introducing new complexities—requires a balanced approach, where ethical considerations are paramount

4. **Cost Efficiency:** On the economic front, the adoption of robotics and automation significantly alters the financial dynamics within manufacturing industries. Initial investments are substantial, but the long-term benefits—such as reduced labor costs, decreased downtime, lower energy consumption, and minimized waste—contribute to a more cost-efficient manufacturing process. These economic benefits align with the principles discussed in the economics of engineering decisions.

   However, the economic benefits must also be weighed against the potential for significant capital expenditures and the need for continuous upgrades and maintenance. The shift towards more automated systems might also require retraining staff to manage and maintain these advanced technologies, introducing both challenges and opportunities for workforce development.

Chapter 14: Instrumentation, Controls, and Measurements

Navigating Complex Control Systems and Dynamic Responses

Understanding Control Systems

Control systems are integral to managing the behavior of dynamic systems in engineering. These systems range from simple feedback loops in home heating systems to sophisticated controllers in automotive and aerospace applications. A control system's primary goal is to regulate the behavior of a machine or process to achieve desired outputs, despite disturbances or changes in the operating environment.

1. **Feedback Control Systems**: Feedback control systems measure a system's output and compare it to the desired setpoint. The system adjusts inputs based on the error to achieve the target output. This type of control is reactive, continuously correcting deviations to maintain stability and accuracy. Common applications include temperature regulation in HVAC systems, speed control in motors, and flight stability in aircraft.

2. **Feedforward Control**: Unlike feedback systems, feedforward control anticipates disturbances by adjusting inputs preemptively. This method requires knowledge of the disturbance's nature and its effects on the system. Feedforward control is proactive, making corrections before disturbances impact the system. It is used in scenarios where disturbances are predictable, such as in chemical process control where changes in raw material properties are known in advance.

Dynamic Response Analysis

The dynamic response of a control system describes how it reacts over time to external inputs. Understanding this response is vital for designing systems that can quickly and efficiently reach their desired state without overshooting or oscillating.

1. **Time Response**: It involves characterizing how a system's output changes over time in response to a standard input, typically a step or impulse. Key metrics include:

    Rise Time: The time it takes for the system's response to go from a lower to a higher threshold, typically from 10% to 90% of the final value.

    Settling Time: The time needed for the system to stay within a specific percentage (typically 2% or 5%) of its final value.

    Overshoot: The extent to which the system exceeds its final steady-state value before settling down.

To understand the concept of thermal resistance, imagine you have a thick wool blanket. When you place the blanket over a hot object, it slows down the heat transfer to your hand. The thicker the blanket, the higher the thermal resistance, and the less heat you feel. Similarly, in a control system, if the system's resistance to change is high, it will react more slowly to external inputs. These indicators help engineers assess system performance and stability.

2. **Frequency Response**: Frequency response analysis determines how a system responds to different input frequencies, focusing on amplitude and phase shifts. This method helps in understanding the system's robustness and identifying potential resonance issues. Engineers use Bode plots and Nyquist plots to visualize frequency response, aiding in the design of systems that can handle a wide range of operating conditions without failure.

Stability Analysis

Many aspects are studied, including dynamical elements, changes, energy transfers, but the most researched point in engineering is precisely stability. That's because a stable control system remains in a consistent state over time or returns to its initial condition after a disturbance. And if we think about it. it is the most coveted condition because it is the best.

Here are some methods that are used:

1. **Root Locus and Nyquist Plots**: These graphical techniques assist in forecasting system stability by depicting the roots of the system's characteristic equation or charting the frequency response. Engineers use these tools to adjust system parameters and ensure stability across all operating conditions.

$$Z = P + N$$

Where Z represents the number of zeros of the closed-loop transfer function in the right-half s-plane (indicating unstable poles), P indicates the number of poles of the open-loop transfer function in the right-half s-plane, and N is the number of encirclements of the critical point by the Nyquist plot of the open-loop transfer function.

2. **Bode Plots**: By representing a system's frequency response with magnitude and phase plots, Bode plots are instrumental in designing feedback loops and tuning controllers to enhance system stability and performance.

Real-World Applications

Control systems are ubiquitous in engineering applications, from robotic arms in manufacturing to autonomous systems in vehicles. Effective control of these systems enhances efficiency, safety, and reliability.

1. **Automotive Systems**: Modern vehicles incorporate complex control systems for engine management, stability control, and autonomous driving features. Understanding the dynamics of these controls is essential for improving vehicle safety and performance.

2. **Aerospace Applications**: Aircraft and spacecraft rely heavily on advanced control systems to manage everything from basic flight controls to intricate maneuvers. Engineers must understand and predict the dynamic responses of these systems under various flight conditions to ensure safety and effectiveness.

3. **Manufacturing**: Automated production lines use control systems to maintain high precision and consistency. The ability to navigate and optimize these systems allows for enhanced productivity and minimal downtime.

Enhancing Precision in Engineering Measurements

Precision in measurements is a necessity; from the design phase to the final product testing, the integrity of a project hinges significantly on the reliability and accuracy of the measurements taken during the process.

Let's clarify fundamentals of measurement accuracy and precision

Accuracy vs. Precision

- **Accuracy** refers to how close a measurement is to the true value.
- **Precision**, on the other hand, indicates the repeatability of measurements, or how close the measurements are to each other, regardless of how close they are to the actual value.

For practical engineering applications, and particularly for your exam preparation, it's vital to grasp that a measurement system can be precise without being accurate and vice versa.

Let's consider a scenario where a scale consistently measures weight 5 grams lighter than the actual weight, it is precise but not accurate. Conversely, a scale that sometimes gives the correct weight and sometimes does not is accurate on occasion but not precise.

Ensuring both accuracy and precision in measurements is achieved through calibration. Calibration involves comparing measurements against known standards and adjusting the measurement system accordingly. Regular calibration is essential to maintain the reliability of measurement systems, thereby ensuring that both accuracy and precision are achieved in engineering practices.

Error Analysis

Measurement errors impact the accuracy and reliability of data. These errors can be broadly classified into two categories:

- **Systematic Errors**: are predictable and typically consistent in magnitude and direction. These arise from identifiable sources, such as instrument calibration issues, environmental conditions, or procedural flaws. Because they are consistent, systematic errors can often be corrected once identified. For instance, let's take the same scale as an example, if it's consistently off by a certain amount, this error can be adjusted through calibration.

- **Random Errors**: in contrast, occur without consistent predictability and vary in magnitude and direction. These errors can result from unpredictable factors like electrical noise, human error, or environmental fluctuations. Random errors are more challenging to correct due to their inherent unpredictability.

Our task is to minimize these errors, and to do so, you can employ several techniques. Careful calibration against known standards is essential here. Using highly reliable and sensitive instruments can also help minimize both systematic and random errors. Techniques such as averaging multiple measurements or using statistical analysis to identify outliers can enhance the overall accuracy and precision of data.

Techniques to Enhance Measurement Precision

1. **Calibration:**
    - Regular calibration against standard references ensures that measurement instruments maintain their accuracy over time. This is crucial for tasks that require high precision, such as component machining and assembly in aerospace engineering.
2. **Advanced Instrumentation:**
    - Utilizing state-of-the-art measurement tools like laser trackers, coordinate measuring machines (CMMs), and digital calipers can significantly enhance the precision of measurements. These tools offer greater resolution and less susceptibility to user error compared to traditional instruments.
3. **Environmental Controls:**
    - Many precise measurements can be influenced by environmental factors like temperature, humidity, and vibration. Controlling these factors can enhance the precision of the

measurements. For example, temperature-controlled environments are used during the calibration of precision instruments.
4. **Statistical Process Control (SPC):**
    - SPC involves applying statistical methods to monitor and control a process, ensuring it operates at its full potential. By measuring and controlling variability, engineers can optimize the quality and precision of their products.

Sensors

Sensors are the backbone of modern instrumentation and control systems, serving as the critical interface between the physical world and the digital realm. They provide essential data about various process parameters, enabling the monitoring, control, and optimization of engineering systems.

There are severale type of sensors, all with different features and situations uses

1. **Temperature Sensors**: These sensors measure the temperature of a system and are crucial in applications ranging from industrial processes to consumer electronics. Common types include:
    - **Thermocouples**: Utilize the Seebeck effect to produce a voltage that correlates with temperature.
    - **RTDs (Resistance Temperature Detectors)**: Utilize the change in a material's resistance with temperature to deliver accurate temperature readings.
    - **Thermistors**: Semiconductors whose resistance changes significantly with temperature, used in applications requiring high sensitivity.
2. **Pressure Sensors**: These sensors measure the force exerted by a fluid or gas on a surface. They are essential in applications such as hydraulic systems, pneumatic controls, and weather stations. Key types include:
    - **Piezoelectric Sensors**: Generate an electric charge in response to pressure changes.
    - **Strain Gauge Sensors**: Measure the change in resistance of a material as it deforms under pressure.
3. **Flow Sensors**: Flow sensors measure the rate of fluid movement through a system, crucial for applications in chemical processing, HVAC, and water treatment. Common types include:
    - **Electromagnetic Flowmeters**: Measure flow using the principle of electromagnetic induction.
    - **Ultrasonic Flowmeters**: Use ultrasonic waves to measure the velocity of the fluid flow.
4. **Level Sensors**: These sensors determine the level of a liquid or solid within a container. They are pivotal in industries such as food and beverage, chemicals, and water treatment. Types include:
    - **Ultrasonic Sensors**: Use sound waves to measure the distance to the surface of the liquid.
    - **Capacitive Sensors**: Measure changes in capacitance caused by the level of the material.

How does it work?

Sensors typically transform a physical quantity into an electrical signal, which is then subject to amplification, conditioning, and processing. Choosing a sensor involves considering factors like measurement range, accuracy requirements, environmental conditions, and application-specific needs.

Where are they used?

Basically everywhere. Sensors are ubiquitous across various fields:

- **Industrial Automation**: Sensors monitor and control manufacturing processes, ensuring quality and efficiency. Temperature, pressure, and flow sensors are commonly used in automated production lines.
- **Healthcare**: in medical devices they monitor patient vital signs, temperature, heart rate and glucose levels, improving diagnosis and treatment.
- **Automotive Systems**: Modern vehicles use a variety of sensors for engine management, safety systems, and autonomous driving. Sensors ensure optimal performance, safety, and compliance with environmental regulations.
- **Environmental Monitoring**: Even here, sensors measure air and water quality, temperature, humidity, and other environmental parameters, aiding in environmental conservation and regulatory compliance.

Right Sensor

Sensor selection is important when parameters are to be measured, and choosing the appropriate one for a given application involves consideration of several factors:

- **Measurement Range and Accuracy**: Ensure the sensor's range and accuracy meet the requirements of the application.
- **Environmental Conditions**: Consider factors such as temperature, humidity, and exposure to chemicals or radiation that might affect sensor performance.
- **Response Time**: For dynamic systems, select sensors with an appropriate response time to ensure accurate and timely data collection.
- **Compatibility**: Ensure the sensor is compatible with the existing instrumentation and control systems.

Importance of Signal Conditioning

Signal conditioning is a fundamental process in data acquisition and control systems because the raw signals that are recorded often require encoding to ensure that they are suitable for further processing and analysis. In other words, it involves manipulating an electrical signal so that it can be analyzed accurately and effectively by digital control systems.

Below are some signal conditioning techniques:

1. **Amplification**: Amplifying a signal involves increasing its voltage level to make it readable for data acquisition systems or to fit within a specific input range. This is especially important for signals that originate as weak outputs from sensors.
2. **Filtering**: Filtering is crucial to eliminate noise and unwanted frequency components from a signal. Low-pass filters, for example, allow signals below a certain frequency to pass through while attenuating frequencies above this threshold, which is vital for reducing high-frequency noise.
3. **Linearization**: Many sensor outputs are non-linear; linearization processes are used to convert these outputs into a linear scale that correlates more directly with the physical measurements.
4. **Isolation**: Electrical isolation separates input and output circuits physically and electrically. It protects devices and personnel from high voltages and reduces noise, ensuring that signals are not contaminated by interference from other equipment.

Analog vs. Digital Signals

Analog signals are continuous and can represent changes over a spectrum of values within certain ranges. These signals are typically generated by natural phenomena, such as sound, light, temperature, and pressure, and can take any value within a given range. For example, a sine wave is a common representation of an analog signal, where its amplitude can vary smoothly over time.

Unlike analog signals, digital signals are not continuous; they are composed of distinct levels or steps. They are discrete and use binary values (0s and 1s) to represent information. Digital signals are less susceptible to noise and interference, making them more reliable for long-distance transmission and data storage. This is because any noise can be filtered out, and the signal can be reconstructed accurately from its discrete values.

Since most modern control systems and computers inherently process digital data, converting analog signals from sensors into digital formats is essential for compatibility and further processing. This conversion is performed using Analog-to-Digital Converters (ADCs), which sample the continuous analog signal at specific intervals and convert these samples into binary data. The precision of this conversion depends on the resolution and sampling rate of the ADC. A higher resolution allows for more precise representation of the analog signal, while a higher sampling rate ensures that the signal is captured more accurately over time.

Basically, analog signals offer a rich, continuous representation of data, ideal for capturing real-world phenomena. Digital signals, by virtue of their discrete nature, provide robustness and accuracy in data transmission and processing, making them indispensable in today's technological landscape.

Analog-to-Digital Converters (ADCs)

As discussed previously, analog signals represent real-world phenomena such as temperature, pressure, sound, and light, in a continuous manner. However, for these analog signals to be processed by digital systems, such as microcontrollers and digital processors, they must be converted into a digital format. This conversion is where ADCs come into play.

They operate by sampling the analog signal at regular intervals and then quantizing these samples into digital values. This process have two steps:

1. **Sampling**: The analog signal is measured at discrete intervals, with the rate of these measurements referred to as the sampling rate. As stated by the Nyquist-Shannon sampling theorem, the sampling rate must be at least twice the highest frequency in the signal to ensure all information is accurately captured.

2. **Quantization**: Each sampled value is then approximated to the nearest value within a finite set of levels. This process converts the continuous amplitude of the analog signal into a digital value. The precision of this quantization is determined by the ADC's resolution, typically measured in bits. A higher resolution ADC can represent the analog signal more accurately, as it divides the signal into a greater number of discrete levels.

ADC Characteristics

- Resolution: This describes the number of distinct values an ADC can output within the range of analog input values. A higher resolution means the ADC can detect finer increments of input signal change.
- Sampling Rate: The rate at which the ADC samples the analog signal. Higher sampling frequencies can capture more rapid changes in the signal, which is crucial for dynamic systems.

Where they are applied?

ADCs find application in a wide range of fields, if we note many in common with the sensors themselves that do not need signal encoding:

- **Industrial Automation**: In automated manufacturing processes, ADCs convert sensor readings (temperature, pressure, flow) into digital data that can be used to control machinery and optimize production lines.

- **Healthcare:** Medical devices such as ECG monitors, blood glucose meters, and imaging equipment rely on ADCs to convert physiological signals into digital data for analysis, diagnosis, and monitoring.

- **Consumer Electronics:** Devices like smartphones, digital cameras, and audio equipment use ADCs to process input from microphones, cameras, and other sensors, enhancing user experience through improved accuracy and performance.

- **Automotive Systems:** Modern vehicles incorporate numerous sensors for engine management, safety systems, and autonomous driving. ADCs convert the analog signals from these sensors into digital data for processing and decision-making.

- **Telecommunications:** ADCs are critical in telecommunications for converting analog voice signals into digital data for transmission over digital communication networks.

Applications in Mechanical Engineering

Manufacturing: In manufacturing, precise measurements are critical to maintaining the integrity of the assembly line and ensuring that the manufactured parts meet design specifications. Techniques such as automated optical inspection and robotic metrology are employed to enhance precision. Automated optical inspection systems use cameras and image processing algorithms to defects and verify dimensions at high speed. Robotic metrology systems, on the other hand, use robotic arms equipped with advanced sensors to perform precise measurements of complex geometries, ensuring that parts are manufactured within tight tolerances.

Quality Assurance: Once again, precision in measurements is key to quality assurance practices. Accurate measurements ensure that components and assemblies function safely and effectively without failures, thereby safeguarding the manufacturer's reputation and minimizing the risk of costly recalls. Quality assurance involves statistical process control (SPC), where data from measurements are analyzed to detect any deviations from the desired specifications. This proactive approach helps in maintaining consistent product quality and reliability.

Research and Development: In R&D, engineers often push the boundaries of what is technically possible, necessitating extremely precise measurements. Innovations in materials science, for example, require measurements at the molecular or even atomic level. Advanced techniques such as atomic force microscopy (AFM) and scanning electron microscopy (SEM) are used to achieve the required precision.

Automotive Engineering: The automotive industry relies heavily on precision measurements for engine components, safety systems, and overall vehicle assembly. Techniques like coordinate measuring machines (CMMs) are used to verify the dimensions of critical parts.

Practice

Explain the difference between feedback and feedforward control systems. Provide an example of each from industrial applications.

Describe how a thermostat in a home heating system uses feedback control to maintain the desired temperature. What signals would be considered as feedback?

For a feedforward control system, provide a scenario in chemical processing where this type of control might be preferred over feedback control. Explain why.

Calculate the rise time and overshoot for a feedback control system where the system's response to a step input is modeled by the function $y(t) = 1 - e^{-2t}$.

Sketch a Bode plot for a system with a gain of 10 dB and a single pole at 1 rad/s. Explain how this plot helps in assessing system stability.

Using the Nyquist stability criterion, determine the stability of a control system whose open-loop transfer function is $G(s) = \frac{10}{s(s+10)}$.

Design a simple feedback loop for a motor speed control system using the PID (Proportional-Integral-Derivative) controller concepts. Explain how each component (P, I, D) contributes to the system's performance.

Given a thermal system where the heat transfer dynamics can be modeled as a first-order process with a time constant of 5 minutes, calculate the system's response time to reach 95% of a step change in temperature input.

Discuss how the Root Locus technique can be used to improve the stability of a control system with a given characteristic equation. Provide a hypothetical equation for illustration.

Explain how sensors play a critical role in automotive safety systems and provide examples of at least two specific applications.

How does the precision of an Analog-to-Digital Converter (ADC) affect the quality of digital signal processing in a sound engineering application?

Compare the effects of environmental conditions on the accuracy of pressure sensors in outdoor applications. Suggest methods to mitigate error.

Describe the process of calibrating a temperature sensor and discuss the potential consequences of poor calibration practices.

Illustrate how the concept of thermal resistance applies to electronic circuit design and identify measures to enhance heat dissipation.

Consider a hydraulic system with a flow sensor. If the sensor has a known systematic error, how can this be corrected through calibration or data processing?

Define and calculate the Re for water flowing at 1 m/s in a pipe with a diameter of 0.5 m, and discuss the flow regime based on your calculation.

Explain the significance of signal conditioning in a noisy industrial environment and provide an example of how filtering could be used to improve signal quality.

Describe the importance of statistical process control (SPC) in a manufacturing setting where precision is critical, such as in semiconductor manufacturing.

Calculate the output of a PID controller given that the setpoint is 100 units, the current process variable is 90 units, the proportional gain is 2, the integral gain is 1, and the derivative gain is 0.5. Assume no previous error has accumulated and that the rate of change of error is constant at -2 units per second.

Propose a setup using a strain gauge sensor to monitor the structural integrity of a bridge and discuss how the data from this sensor might be used in real-time safety assessments.

Chapter 15: Design and Analysis for Mechanical Reliability

Mechanical Design Principles and Stress Analysis

In mechanical engineering, design principles form the backbone of creating functional, durable, and reliable machinery and structures. These principles guide engineers in developing products that meet specified requirements while optimizing performance and minimizing costs. Effective mechanical design hinges on a deep understanding of materials, mechanics, and the anticipated loads and stresses a product will encounter during its lifecycle.

1. **Design for Reliability**: The foremost goal in mechanical design is ensuring reliability. This involves considering all possible modes of failure and designing to mitigate them. We've discussed the importance of material properties and load analysis in previous chapters. Expanding upon this foundation, the design for reliability also incorporates a variety of robust engineering practices aimed at optimizing product life and performance. This includes the application of safety factors, the use of redundant systems, and the implementation of fail-safe mechanisms.

2. **Load and Stress Analysis**: Every mechanical component in a system must be capable of withstanding the forces and stresses exerted upon it during operation. Stress analysis, a critical aspect of mechanical design, involves calculating the stresses and deformations that will occur in a material under given loads. This analysis helps in identifying weak points that might fail under stress.

    - **Static Stress Analysis**: Examines the stress response of a material or structure under static (constant) loading conditions. This type of analysis ensures that a component will not fail under prescribed loads and is fundamental for structural integrity assessments.

    - **Dynamic Stress Analysis**: On the other hand, looks at how materials and structures respond to varying loads over time, such as those experienced in automotive and aerospace applications, where the conditions of use can change rapidly and unpredictably.

3. **Factor of Safety (FOS)**: Integral to mechanical design, the Factor of Safety is a design criterion that provides a safety margin in uncertain conditions. It is defined as the ratio of the maximum stress a material can withstand to the designed allowable stress. Incorporating an appropriate FOS ensures that even in the worst-case scenario, the system remains safe and operational.

4. **Simplification and Optimization**: While designing, it's crucial to simplify the design as much as possible without compromising on functionality. This not only reduces the cost of production but also eases the maintenance and repair processes. Optimization techniques, such as finite element analysis (FEA), are employed to refine the design further, enhancing performance while reducing material usage and weight.

Reliability Prediction

Reliability prediction and estimation techniques are vital tools for quantitatively assessing the expected performance and longevity of mechanical systems based on historical data and testing. As we've touched upon in earlier discussions about quality and performance, reliability estimation integrates seamlessly into these themes, providing a quantitative backbone to design decisions and maintenance planning.

Techniques for Reliability Prediction

1. **Mean Time Between Failures (MTBF)**:
    - **Definition**: MTBF is a statistical measure used to predict the time interval between inherent failures of mechanical or electronic systems during operation. It is commonly used to estimate and compare the reliability of different systems.
    - **Application**: MTBF can planning maintenance schedules, designing redundancy into systems, and enhancing overall system design to minimize the likelihood of failures.

2. **Reliability Block Diagrams (RBD)**:
    - **Definition**: RBDs are graphical representations that illustrate the interconnections and dependencies of system components from a reliability perspective. Each block in the diagram represents a system component, and the layout reflects how components contribute to the system's overall reliability.
    - **Utility**: RBDs help to visualize the paths through which a system can fail and evaluate the impact of individual component failures on the entire system's functionality.

3. **Statistical Reliability Models**:
    - **Concepts**: These models use statistical methods to analyze historical failure data and testing results to predict future reliability. Common models include the Exponential, Weibull, and Log-Normal distributions, each appropriate under different conditions based on the nature of the data and failure mechanisms involved.
    - **Implementation**: are used to estimate failure probabilities, assess risks, and determine necessary improvements or changes in materials and designs to meet reliability targets.

Enhancing Reliability Through Data and Testing

We've learned that the practical application of reliability techniques involves gathering and examining extensive data from both real-world operations and controlled experiments. Here's a detailed look at how these techniques are generally implemented:

1. **Data Collection:**
    - Accurate and comprehensive data collection is crucial for reliability analysis. This involves gathering data from operational history, including the environments of use, loading conditions, and instances of failure. Data sources may include sensors, logs, maintenance records, and user feedback.
    - **Operational History**: Collecting data from equipment in its working environment provides insights into the actual conditions under which it operates, which can differ significantly from controlled testing environments.
    - **Loading Conditions**: Understanding the specific loads and stresses a component experiences during operation helps in identifying potential failure modes.
    - **Failure Instances**: Recording details about when and how failures occur allows for better understanding and mitigation of these issues in future designs.
2. **Model Fitting and Analysis:**
    - Once data is collected, statistical models are applied to identify trends, predict failures, and calculate reliability metrics such as Mean Time Between Failures (MTBF).
    - **Trend Identification**: Analyzing data to find patterns and correlations helps in understanding the underlying causes of failures.
    - **Failure Prediction**: Using models to estimate the likelihood and timing of future failures enables proactive maintenance and design improvements.
    - **Reliability Metrics**: Metrics like MTBF provide a quantifiable measure of system reliability, which is essential for planning and decision-making.
3. **Iterative Improvements:**
    - Insights gained from statistical analysis and model predictions are used to make iterative improvements to design, material selection, and maintenance strategies.
    - **Design Improvements**: Refining design elements based on data can enhance durability and performance. This might involve changing materials, redesigning components, or adding redundancies.
    - **Material Selection**: Choosing more suitable materials based on their performance in actual conditions can reduce the incidence of failures.
    - **Maintenance Strategies**: Developing maintenance schedules and strategies based on predicted failure times can prevent unexpected downtimes and extend the life of the equipment.

Environmental Stress Screening (ESS)

ESS is a critical process used to improve the reliability of electronic components and systems. By intentionally subjecting these components to various environmental stresses, it helps identify and eliminate potential defects that could cause failures during normal operation. This proactive approach ensures that only the most robust components make it to the final product, significantly improving overall reliability.

The practical application of reliability techniques involves the collection and analysis of large amounts of data from both real operations and controlled testing. Here is how SSE fits into this process:

- Data collection:

SSE plays a key role in the initial phase of data collection; in fact, data are collected on how components respond to various stress conditions. This data comes from sensors, logs, maintenance records, and user feedback, providing a solid basis for reliability analysis.

- Model fitting and analysis:

Once the data is collected, it is analyzed to identify trends, predict failures, and calculate reliability metrics such as average system uptime.

Techniques and Procedures in ESS

1. **Temperature Cycling**: Products are exposed to extreme temperature variations that they might encounter in real-world operations. This helps in identifying material and structural weaknesses that could lead to failures due to thermal expansion and contraction.
2. **Vibration Testing**: Simulates the conditions that products might face during shipping and handling or operational movements. This test is crucial for detecting mechanical and structural weaknesses that can lead to operational failures.
3. **Humidity Testing**: Exposes the product to high levels of humidity to test for moisture ingress and its effects on the product. This is particularly important for electronic and metallic components susceptible to corrosion or short-circuiting.
4. **Altitude Testing**: Products are tested under low-pressure conditions that might be experienced at high altitudes.

Goals and Outcomes of ESS

The primary objective of Environmental Stress Screening (ESS) is to significantly boost product reliability by pinpointing and rectifying potential failures at an early stage. Through rigorous testing under controlled, extreme environmental conditions, ESS identifies both defects and design vulnerabilities. This enables engineers to implement essential modifications during the developmental phases, well before the commencement of full-scale production. Moreover it minimizes occurrences of failures in the field, reduces warranty-related expenditures, and substantially improves customer satisfaction by guaranteeing that the products are not only robust but also exceptionally reliable in diverse operating conditions.Integration of ESS in Design and Production

Implementing ESS requires careful planning and integration into the design and production processes:

- **Design Phase**: ESS should be considered during the initial design phase. Designers need to anticipate the types of stress tests that the product will undergo and consider these when choosing materials and structural layouts.
- **Prototype Testing**: ESS is typically applied during the prototype testing phase. Results from these tests inform further design adjustments and refinements to enhance durability and performance.
- **Production Quality Control**: Incorporating ESS as a routine part of the quality control process in production ensures that each batch of products meets stringent reliability standards before reaching the market.

Case Studies in Mechanical Design

In mechanical engineering, design principles form the backbone of creating functional, durable, and reliable machinery and structures. These principles guide engineers in developing products that meet specified requirements while optimizing performance and minimizing costs. Effective mechanical design hinges on a deep understanding of materials, mechanics, and the anticipated loads and stresses a product will encounter during its lifecycle.

Problem Definition: To design a gear capable of withstanding high torque and thermal stresses typical of an automotive transmission system without failing prematurely, is the goal. The design must ensure durability and reliable performance under these demanding conditions.

Material Selection: Based on the operating conditions, a high-strength steel alloy is chosen for its excellent fatigue resistance and superior thermal properties. This selection ensures the gear can handle the cyclical loads and elevated temperatures encountered during operation.

Stress Analysis: Using Finite Element Analysis (FEA), the gear's stress distribution under operational loads is thoroughly examined. This process involves creating a detailed model of the gear and applying expected loads to identify areas susceptible to high stress concentrations. These critical areas are then targeted for redesign or reinforcement to prevent potential failure.

Design Optimization: The gear teeth are optimized for shape and size to ensure even stress distribution. Additionally, lubrication channels are integrated into the design to enhance heat dissipation and reduce thermal stress, further improving the gear's performance and longevity.

Prototype Testing: A prototype gear, manufactured based on the optimized design, undergoes rigorous testing under actual operating conditions. This testing phase validates the design assumptions and assesses the effectiveness of stress reduction strategies implemented in the design.

Iteration and Finalization: Feedback from the prototype testing informs further refinements to the gear design. Adjustments are made to enhance reliability and performance, ensuring the final product meets all operational requirements and exceeds durability expectations.

Ensuring Quality and Reliability

In the field of engineering, the quality and reliability of designs are paramount. We have repeated this many, many times. We will continue to do so because it is a fundamental concept and as such must become your mantra .

Integration of Quality Management Systems

Quality management systems (QMS) are of utmost importance. They provide a structured framework that helps engineers adhere to international standards and meet customer requirements consistently. Implementing a QMS like ISO 9001 can guide the entire design process, from initial concept through manufacturing and deployment, ensuring that quality and reliability are baked into the product at every stage.

1. **Document Control**: Maintaining rigorous documentation processes ensures that every design change, decision, and test result is recorded.
2. **Standardization**: Adopting industry standards ensures that parts and processes meet universal quality criteria, facilitating integration and compatibility across different systems and ensuring reliability.

Reliability Engineering Techniques

Reliability engineering is focused on ensuring a product performs as expected without failure for a specified period within its intended environment. Several techniques are instrumental in achieving this:

Failure Modes and Effects Analysis (FMEA): FMEA is a methodical process employed to detect potential failure modes in a product or process, to understand the risk associated with those failures, and to implement measures to mitigate those risks. It helps engineers anticipate and prevent potential failures, enhancing product reliability.

Steps for FMEA:

1. **Define the Scope and Assemble a Team**:

    - **Scope Definition**: Clearly define what part of the system you are analyzing.
    - **Team Assembly**: Form a team with diverse expertise relevant to the product or process.

2. **Identify Potential Failure Modes**:

    - **Brainstorming**: Conduct sessions to list all possible ways something can fail.
    - **Historical Data**: Look at past records to find common failures.

3. **Determine the Effects of Each Failure Mode**:

    - **Impact Analysis**: Assess how each failure affects the system, focusing on safety, performance, and customer impact.
    - **Severity Rating**: Rate the impact from 1 (no effect) to 10 (catastrophic effect).

4. **Identify the Causes of Each Failure Mode**:

    - **Root Cause Analysis**: Find out the root causes of each failure.
    - **Occurrence Rating**: Rate the likelihood of each cause from 1 (unlikely) to 10 (very likely).

5. **Identify Current Controls and Detection Methods**:

    - **Control Review**: List existing controls and methods to detect failures.
    - **Detection Rating**: Rate how effective these controls are from 1 (highly effective) to 10 (ineffective).

6. **Calculate the Risk Priority Number (RPN)**:

    - **RPN Calculation**: Use the formula:

    $$RPN = Severity \times Occurrence \times Detection$$

    - **Prioritization**: Focus on the failure modes with the highest RPNs.

7. **Develop and Implement Mitigation Strategies**:

    - **Action Plan**: Create a plan to address high-priority failure modes, which might include redesigning parts or improving materials.
    - **Implementation**: Apply the mitigation strategies and monitor their effectiveness.

8. **Document and Review**:

    - **Documentation**: Record all findings and actions taken during the FMEA process.
    - **Continuous Improvement**: Regularly update the FMEA with new information or changes to the product or process.

1. Fault Tree Analysis (FTA): FTA is a deductive, top-down method used to analyze the causes of system level failures. This technique helps identify the root causes of failures and their interrelationships, which is crucial for developing strategies to reduce the likelihood of failure.

Statistical Tools for Quality and Reliability

Statistical tools are essential in assessing the quality and reliability of engineering designs:

1. **Statistical Quality Control (SQC)**: SQC uses statistical methods to monitor and control manufacturing processes. Methods like control charts and process capability analysis aid in ensuring that the process runs efficiently, resulting in more products that meet specifications and less waste from rework or scrap.
2. **Reliability Testing**: This includes life testing and accelerated life testing, where products are subjected to stress and strain beyond their normal service levels to identify potential modes of failure.

Steps for Implementing SQC

1. **Monitoring and Controlling Processes**:

   Control Charts: These are used to track data over time and detect any variations that may indicate problems.

   - **Types of Control Charts**:
     - **X-Bar and R Charts**: Track the average and range of a sample.
     - **P Charts**: Monitor the proportion of defective items in a process.

2. **Interpreting Control Charts**: Identify patterns and trends to determine if a process is in control or if there are variations that need attention.

Process Capability Analysis:

Key Metrics:

- **Cp (Process Capability Index)**: Measures how well a process can produce output within specification limits. Higher values indicate a better process.
- **Cpk (Process Capability Performance Index)**: Similar to Cp, but also considers how centered the process is within the specification limits.

$$Cp = \frac{USL - LSL}{6\sigma}$$

Calculating Cp and Cpk:

$$Cpk = \min\left(\frac{USL - \mu}{3\sigma}, \frac{\mu - LSL}{3\sigma}\right)$$

where USL is the upper specification limit, LSL is the lower specification limit, μ is the process mean, and σ is the standard deviation.

Ensuring Process Efficiency

- **Reducing Waste**: By using control charts and capability analysis, identify areas where the process can be improved to reduce rework or scrap.
- **Continuous Monitoring**: Regularly update and review control charts and process capability metrics to maintain process efficiency and product quality.

This method helps maintain consistent product quality, it reduces waste and rework, saving time and materials. Summary it ensures that processes are stable and capable, leading to better overall performance.

Life Testing

Life Testing involves subjecting a product to normal operating conditions to determine its lifespan. This test helps in understanding how a product performs over time and under typical usage. Key aspects of life testing include:

- **Duration**: The product is tested for a long period to simulate its entire expected lifespan.
- **Monitoring**: Continuous monitoring is done to observe any failures or performance degradation.
- **Data Analysis**: Collected data is analyzed to estimate the product's reliability and to identify common failure modes.

Accelerated Life Testing

There is another way to quantify duration, and this is called the Accelerated Life Testing. It complains to higher stress levels than normal to induce failures more quickly. This approach helps in identifying potential failure modes within a shorter time frame. Key elements include:

- **Stress Levels**: Products are exposed to elevated temperatures, increased loads, or higher voltages to accelerate aging and wear.
- **Failure Analysis**: The failures observed during accelerated testing are analyzed to understand the causes and mechanisms.
- **Extrapolation**: The data obtained is used to predict the product's lifespan under normal operating conditions.

Incorporating Quality by Design (QbD)

Quality by Design (QbD) is an approach that emphasizes planning for quality from the earliest stages of product development. By considering potential reliability issues early on, engineers can create products that are inherently more robust and dependable. Key principles include:

- **Design Reviews**: Regular reviews to ensure that quality and reliability are integrated into every stage of the design process. These reviews involve cross-functional teams evaluating the design for potential issues and improvements.
- **Prototyping and Iteration**: Developing prototypes allows for real-world testing and refinement of designs. Iterative testing and modification based on trial results ensure that the product meets design specifications and adheres to quality and reliability standards.
- **Risk Assessment**: Identifying potential risks and implementing strategies to mitigate them is a core component of QbD. This involves using tools like FMEA and FTA to anticipate and address potential failure modes.

Practice

Define Mechanical Reliability. What is the primary goal of incorporating reliability considerations in mechanical design?

Calculate Factor of Safety (FOS): Given a component with a maximum allowable stress of 120 MPa and an expected working stress of 80 MPa, calculate the Factor of Safety.

Static Stress Analysis: Determine the highest bending stress in a beam that spans 10 meters and experiences a uniform load of 3000 N/m, assuming it has simple supports at each end.

Dynamic Stress Analysis: Describe how you would model the dynamic stress on an automotive suspension component using finite element analysis during a typical road bump impact.

MTBF Calculation: If a system has three components in series with MTBFs of 100, 200, and 300 hours respectively, calculate the overall MTBF of the system.

Reliability Block Diagrams (RBD): Sketch a reliability block diagram for a system comprising two parallel components each with a reliability of 0.95.

Mean Time Between Failures (MTBF) vs. Reliability: Explain how MTBF is related to reliability in a practical engineering scenario.

Design for Reliability: Propose a design modification to improve the reliability of a heat exchanger prone to fouling.

Statistical Reliability Model Application: Compute the likelihood that a component, which has an exponential failure rate distribution with an average failure time of 500 hours, will last at least 700 hours without failing.

Environmental Stress Screening (ESS): List three environmental stresses that would be critical to test for a smartphone's durability.

Temperature Cycling Test: Explain the significance of temperature cycling in Environmental Stress Screening and its impact on component reliability.

Vibration Testing in ESS: Design a simple vibration test protocol for a new aircraft component to identify potential failure modes under operational stress.

Humidity Testing: Discuss how you would set up a test to evaluate the effects of humidity on the corrosion of electronic circuit boards.

Failure Modes and Effects Analysis (FMEA): Conduct a basic FMEA for a bicycle brake system focusing on the brake cable.

Fault Tree Analysis (FTA): Draw a simple fault tree that outlines the potential causes of failure for an overheating engine.

Bode Plot Interpretation: Given a Bode plot of a control system, determine the system's stability margins and discuss potential stability improvements.

Nyquist Plot Stability Assessment: Analyze a Nyquist plot to determine if a closed-loop system is stable based on the number of encirclements of the critical point.

Load and Stress Analysis Problem: Calculate the shear stress on a bolt used in a flanged connection under a tensile load of 10 kN.

Root Locus Plot Analysis: Explain how you would use a root locus plot to design a feedback controller for a motor speed control system.

Reliability Enhancement Strategy: Propose a strategy to enhance the reliability of a water pump used in a remote location, considering both design and operational parameters.

Chapter 16: Practical Exam Preparation: Simulated Tests and Solutions

Full-Length Practice Exams Tailored to the FE Specifications

1EXAM PREP.

1. Calculate the radius and the center coordinates of a circle that passes through the points (1,1), (2,4), and (5,3).

A) Center: (3, 2), Radius: $\sqrt{5}$
B) Center: (2, 1), Radius: 5
C) Center: (3, 2), Radius: 2
D) Center: (4, 2), Radius: 2.5

2. Determine the maximum and minimum values of the function $f(x) = x^3 - 6x^2 + 9x + 15$ within the interval [0,5].

A) Minimun: 20, Maximum: 25
B) Minimum: 15, Maximum: 35
C) Minimum: 15, Maximum: 38
D) Minimum: 3, Maximum: 22

3. Find the critical points and classify them for the function $f(x,y) = x^2 + y^2 - 2x - 6y$.

4. Calculate the area under the curve $y = 4x - x^2$ from $x = 0$ to $x = 4$.

5. Find the inverse of the matrix $\begin{pmatrix} 1 & 2 \\ 3 & 7 \end{pmatrix}$

A) $\begin{pmatrix} 7 & -2 \\ -3 & 1 \end{pmatrix}$

B) $\begin{pmatrix} 7 & -2 \\ 3 & -1 \end{pmatrix}$

C) $\begin{pmatrix} -7 & 2 \\ 3 & -1 \end{pmatrix}$

D) $\begin{pmatrix} -7 & 2 \\ -3 & 1 \end{pmatrix}$

6. Calculate the dot product of vectors $\vec{a} = \langle 3, 4, 5 \rangle$ and $\vec{b} = \langle 2, -1, 3 \rangle$.

7. Approximate the value of $\sqrt{50}$ using the Taylor series expansion of $\sqrt{x}$ around x=49.

   A) 7.07
   B) 7.14
   C) 7.08
   D) 7.10

8. A factory produces light bulbs, and it is known that 5% of them are defective. If you randomly select 20 light bulbs, what is the probability that exactly 2 of them are defective?

   A) 0.26
   B) 0.18
   C) 0.29
   D) 0.12

9. A dataset contains the values {2, 4, 4, 4, 5, 5, 7, 9}. Select all the correct statements about this dataset.

   A) The mean is 5.
   B) The median is 4.5.
   C) The mode is 4.
   D) The standard deviation is 2.

10. Find the 95% confidence interval for the mean of a sample with a mean of 50, a standard deviation of 5, and a sample size of 100.

11. A fair six-sided die is rolled twice. What is the probability that the sum of the two rolls is 7?

    A) $\frac{1}{36}$

    B) $\frac{1}{12}$

    C) $\frac{1}{6}$

    D) $\frac{5}{36}$

12. Select all the true statements about the normal distribution:

    A) It is symmetric about the mean.
    B) The mean, median, and mode are equal.
    C) It has a bell-shaped curve.
    D) Approximately 68% of the data falls within one standard deviation of the mean.

13. According to the NSPE Code of Ethics, which of the following should be the highest priority for engineers?

    A) Employer's interests
    B) Client's interests
    C) Public safety
    D) Professional development

14. An engineer discovers a critical safety issue in a project after it has been completed. What is the most ethical action to take?

    A) Ignore the issue to avoid conflict
    B) Report the issue to the appropriate authorities
    C) Fix the issue quietly without informing anyone
    D) Blame another team member

15. Select all the scenarios that would constitute a conflict of interest for an engineer:

    A) Accepting a gift from a supplier
    B) Owning stock in a competitor's company
    C) Having a close relative working in a client's organization
    D) Volunteering for a professional society

16. Fill in the blank: Engineers must provide services solely within their areas of expertise and should only accept assignments when qualified by _____.

17. An engineer is asked to sign off on a project they did not supervise. According to the code of ethics, what should the engineer do?

    A) Sign off without checking
    B) Review the project thoroughly before signing
    C) Refuse to sign
    D) Ask a colleague to sign on their behalf

18. A project requires an initial investment of $100,000 and is projected to generate annual cash flows of $20,000 for the next 7 years. What is the payback period?

    A) 4 years
    B) 5 years
    C) 3 years
    D) 6 years

19. Select all true statements regarding the time value of money:

    A) The future value is the present value compounded at the interest rate over a period of time

    B) The present value is the future value discounted at the interest rate over time.

    C) Annuities consist of a series of equal payments made at regular intervals.

    D) The interest rate does not influence present value calculations.

20. Determine the present value of $50,000 to be received in 10 years, given an annual discount rate of 5%.

21. A machine is priced at $10,000 and has a useful lifespan of 5 years with no salvage value. If the machine generates $3,000 annually, what is the return on investment (ROI)?

    A) 30%
    B) 50%
    C) 60%
    D) 70%

22. Fill in the blank: The break-even point is reached when the total revenue equals the total _____.

23. A circuit contains a 10-ohm resistor in series with a 5-ohm resistor and a 20V power supply. What is the current flowing through the circuit?

    A) 0.5 A
    B) 1.33
    C) 1.5 A
    D) 2 A

24. Select all the correct statements about alternating current (AC):

    A) The frequency of AC is measured in Hertz.
    B) AC voltage varies sinusoidally with time.
    C) AC is used for long-distance power transmission.
    D) AC has a constant polarity.

25. Calculate the total capacitance of two capacitors, 6 μF and 12 μF, connected in series:
26. In a parallel circuit with two branches, one branch has a resistance of 8 ohms and the other branch has a resistance of 12 ohms. What is the total resistance of the circuit?

A) 8.0 ohms
B) 10.0 ohms
C) 4.8 ohms
D) 6.0 ohms

27. Select all the true statements about magnetic flux:

    A) Magnetic flux is measured in Weber.
    B) Magnetic flux is the product of the magnetic field and the area perpendicular to the field.
    C) Magnetic flux density is also known as magnetic field strength.
    D) Magnetic flux through a closed surface is zero according to Gauss's law for magnetism.

28. Fill in the blank: The inductance of a coil is defined as the ratio of _____ to the rate of change of current.

29. Determine the necessary force at point A to keep a beam in horizontal equilibrium which is hinged at one end and has a uniform distributed load of 200 N/m over its 5 m length.

    A) 250 N

    B) 500 N

    C) 1000 N

    D) 1500 N

30. Select all the true statements about the centroid of composite shapes:

    A) The centroid of a composite shape is the weighted average of the centroids of its parts.
    B) The centroid always lies within the material of the shape.
    C) For symmetric shapes, the centroid lies along the axis of symmetry.
    D) The centroid can be outside the physical boundaries of the shape.

31. Determine the moment of inertia for a rectangle about its base, given that the base measures 300 mm and the height is 500 mm.

32. Calculate the maximum load that can be placed on a 30° inclined plane without slipping, given that the coefficient of static friction is 0.4.

    A) 100 N
    B) 200 N
    C) 300 N
    D) 400 N

33. ☐ Fill in the blank: The tension in a cable holding a 400 kg mass in equilibrium, angled at 45° to the horizontal is _____ N.

34. Determine the shear force at a point 2 m from the left support in a 6 m long simply supported beam with a 500 N point load at the center.

35. Calculate the bending moment at the midpoint of a 4 m long simply supported beam with a uniformly distributed load of 300 N/m.

36. Assess the stability of a 3 m tall column fixed at both ends, subjected to a lateral load of 100 N at the top.

    A) Stable
    B) Unstable
    C) Needs more data for assessment
    D) Depends on the material properties

37. Calculate the forces in all members of a simple truss with three members forming a triangle, subjected to a load of 1000 N at one joint.

38. Determine the force required to maintain equilibrium in a pulley system with a mass of 200 kg hanging from one side.

    A) 1000 N
    B) 1500 N
    C) 2000 N
    D) 2500 N

39. Find the centroid of an L-shaped object formed by two rectangles, one 100 mm by 200 mm and the other 200 mm by 100 mm, overlapping by 50 mm.

40. What is the acceleration of a car that uniformly accelerates from rest to a speed of 20 m/s in 5 seconds?

    A) 2 m/s²
    B) 4 m/s²
    C) 6 m/s²
    D) 8 m/s²

41. Select all the true statements about simple harmonic motion (SHM):

    A) The motion is sinusoidal in time.
    B) The restoring force is proportional to the displacement.
    C) The total energy is constant.
    D) The frequency depends on the amplitude.

42. Determine the natural frequency of a spring-mass system where the mass is 2 kg and the spring constant is 50 N/m.

43. What is the maximum height achieved by a projectile launched at an angle of 30° with an initial velocity of 40 m/s?

    A) 10 m
    B) 30 m
    C) 40 m
    D) 20 m

44. Fill in the blank: The period of a pendulum is influenced by the length of the pendulum and the _____.

45. Determine the velocity of a 5 kg object after it has fallen from rest for 3 seconds under the influence of gravity (g = 9.81 m/s²).

46. Select all the factors that affect the damping ratio in a vibrating system:

    A) Mass of the system
    B) External force applied
    C) Damping coefficient
    D) Stiffness of the system

47. What is the resulting acceleration of a 10 kg object subjected to a force of 100 N?

    A) 5 m/s²
    B) 7 m/s²
    C) 10 m/s²
    D) 15 m/s²

48. Fill in the blank: In a rotating system, the moment of inertia quantifies the resistance to _____.

49. Calculate the kinetic energy of a 3 kg object moving at a speed of 10 m/s.

50. Select all the true statements about the impulse-momentum theorem:

    A) Impulse is calculated by multiplying force and time.

    B) The change in momentum is equal to the impulse applied.

    C) Impulse can be calculated as the area under a force-time graph.
    D) Impulse depends only on the initial velocity.

51. Calculate the frequency of a vibrating string that is 2 m long and has a wave speed of 100 m/s.

52. Calculate the normal stress in a material under a point load of 1000 N with a cross-sectional area of 0.005 m².

    A) 20,000 Pa
    B) 50,000 Pa
    C) 100,000 Pa
    D) 200,000 Pa

53. Select all the correct statements about strain under axial load:
    A) Strain is dimensionless.
    B) Strain is directly proportional to the applied stress.
    C) Strain is inversely proportional to the material's Young's modulus.
    D) Strain is measured in Pascals.

54. Using Hooke's Law, calculate the strain for a material with a stress of 200,000 Pa and Young's modulus of 200 GPa.

55. Calculate the elongation of a steel rod subjected to a force of 2000 N, length of 2 m, cross-sectional area of 0.01 m², and Young's modulus of 210 GPa.

56. Determine the critical buckling load for a column using Euler's formula:
    A) 8,829 kN
    B) 17,657 kN
    C) 35,314 kN
    D) 70,628 kN
57. Calculate Young's modulus for a material with a stress of 250,000 Pa and a strain of 0.0125.
58. Calculate the shear stress in a beam with a shear force of 5000 N, the first moment of area Q = 1000 mm³, and the moment of inertia I = 200 mm^4 over a width t = 50 mm.
59. Determine the maximum deflection of a simply supported beam under a uniformly distributed load
60. Determine the critical speed of a rotating shaft that is 2 meters long and has a diameter of 50 mm. The shaft is made of steel with a density of 7850 kg/m³ and a Young's modulus of 210 GPa. The shaft is supported in a fixed-free configuration. Calculate the critical speed and discuss the factors that might influence this speed.
61. Calculate the stress concentration factor given a maximum stress of 300,000 Pa and nominal stress of 200,000 Pa.
62. For a beam supported at three points with a center load, use compatibility conditions and superposition to find the reactions at each support.
63. Which of the following materials typically exhibits the highest electrical conductivity?

    A) Copper
    B) Aluminum
    C) Steel
    D) Titanium
64. Select all the true statements about phase diagrams:

    A) They show the equilibrium phases of a material at different temperatures and compositions.
    B) They can be used to predict the mechanical properties of alloys.
    C) They are only applicable to pure metals.
    D) Eutectic points indicate the lowest melting point of a mixture.
65. Fill in the blank: The primary strengthening mechanism in steels is the formation of _____.

66. What is the main advantage of using composite materials over traditional metals?

   A) Higher thermal conductivity
   B) Greater strength-to-weight ratio
   C) Lower cost
   D) Easier manufacturing

67. Select all the processing techniques that involve changing the shape of a material:

   A) Casting
   B) Extrusion
   C) Annealing
   D) Forging

68. Fill in the blank: The heat treatment process that involves heating a material and then slowly cooling it to remove internal stresses is called _____.

69. Which property measures a material's ability to resist deformation under load?

   A) Toughness
   B) Hardness
   C) Ductility
   D) Elasticity

70. Select all the effects of alloying elements in steel:

   A) Increased hardness
   B) Improved corrosion resistance
   C) Reduced melting point
   D) Enhanced ductility

71. Calculate the force required to maintain steady flow in a horizontal pipe with a diameter of 0.5 meters if the flow rate is 2 m³/s and the fluid density is 1000 kg/m³.

   A) 1000 N
   B) 2000 N
   C) 20400N
   D) 8000 N

72. Select all the true statements about Reynolds number:

   A) It is a dimensionless quantity.
   B) It determines whether flow is laminar or turbulent.
   C) It is influenced by the fluid's velocity and viscosity.
   D) It is always greater than 4000 for turbulent flow.

73. Fill in the blank: The buoyant force acting on a submerged object is equal to the _____ of the displaced fluid.

74. Determine the pressure drop per meter in a pipe with a dynamic viscosity of 0.001 Pa·s, where water is flowing at a rate of 1 m³/s through a pipe with an internal diameter of 1 meter. Assume the water has a density of 1000 kg/m³ and the kinematic viscosity is 1 x 10⁻⁶ m²/s. Calculate the Reynolds number to identify whether the flow is laminar or turbulent, and use the appropriate formula to calculate the friction factor based on the flow regime. Then, calculate the pressure drop per meter using the Darcy-Weisbach equation.

75. Compute the velocity of water exiting a tank through a hole at its base 10 m below the surface.

    A) 10 m/s
    B) 14 m/s
    C) 20 m/s
    D) 25 m/s

76. Calculate the Reynolds number for water flowing through a pipe with a diameter of 0.05 m at a velocity of 3 m/s, given that the kinematic viscosity is 1 x10⁻⁶ m²/s.

77. Calculate the critical depth for a channel with a flow rate of 5 m³/s and a width of 2 m.

78. Determine the lift force on an airplane wing with an area of 50 m² experiencing a lift coefficient of 0.8 at an airspeed of 250 m/s and air density of 1.225 kg/m³.

79. Fill in the blank: The drag force experienced by an object moving through a fluid is proportional to the _____ of the fluid.

80. Assess the flow rate necessary to achieve a Reynolds number of 4000 in a pipe with a diameter of 0.1 meters and kinematic viscosity of 1.0 x10⁻⁶ m²/s.

81. Calculate the drag force on a car with a frontal area of 2.2 m², drag coefficient of 0.32, and traveling at 27 m/s in air with a density of 1.2 kg/m³.

82. Determine the pressure at a depth of 5 meters in freshwater with a density of 1000 kg/m³.

    A) 49 kPa
    B) 50 kPa
    C) 51 kPa
    D) 52 kPa

83. Which of the following laws asserts that energy cannot be created or destroyed, but only changed from one form to another?

    A) 0 Law
    B) 1 Law
    C) 2 Law
    D) 3 Law

84. ☐ Select all the true statements about entropy:

    A) Entropy measures the disorder of a system.
    B) Entropy increases in irreversible processes.
    C) Entropy is conserved in all processes.
    D) Entropy can decrease in a closed system.

85. Fill in the blank: The _____ cycle is the ideal cycle for a heat engine operating between two thermal reservoirs.

86. Determine the work done by a gas that expands isothermally from a volume of 1 m³ to 3 m³ at a constant temperature, with an initial pressure of 100 kPa.

    A) 69.3 kJ
    B) 79.3 kJ
    C) 89.3 kJ
    D) 99.3 kJ

87. Calculate the change in internal energy of a system that absorbs 500 J of heat and performs 300 J of work.

88. Select all the processes that are isothermal:

    A) Isothermal expansion of an ideal gas

    B) Isothermal compression of an ideal gas

    C) Melting phase change from solid to liquid

    D) Heating water at constant pressure

89. Fill in the blank: The efficiency of a Carnot engine depends on the temperatures of the _____ and _____ reservoirs.

90. Determine the specific heat capacity of a substance if 2000 J of heat is needed to increase the temperature of 1 kg of the substance by 5°C.

    A) 400 J/kg·°C
    B) 200 J/kg·°C
    C) 300 J/kg·°C
    D) 500 J/kg·°C

91. Calculate the heat absorbed by 2 kg of water when it is heated from 20°C to 80°C. The specific heat capacity of water is 4.18 kJ/kg·°C.

92. Select all the true statements about the Second Law of Thermodynamics:

    A) Heat transfers spontaneously from warmer to cooler bodies.

    B) The entropy of an isolated system will never decrease.

    C) Machines that produce perpetual motion of the second kind are feasible.

    D) Converting all heat into work without any losses is impossible.

93. Fill in the blank: The _____ principle states that the entropy of the universe tends to increase.

94. What is the maximum theoretical coefficient of performance (COP) for a refrigerator that operates between 0°C and 30°C?

    A) 9.1
    B) 7.8
    C) 8.8
    D) 9.8

95. Identify the mode of heat transfer that does not need a medium.

    A) Convection

    B) Conduction and Convection

    C) Radiation

    D) Conduction

96. Select all the true statements about thermal conductivity:

    A) It quantifies a material's capacity to conduct heat.

    B) Metals generally have high thermal conductivity.
    C) It is independent of the material's temperature.
    D) Insulators have low thermal conductivity.

97. Fill in the blank: The heat transfer rate through a solid is governed by _____ law of heat conduction.

98. Calculate the heat transfer through a 0.1 m² area wall with a thickness of 0.05 m and thermal conductivity of 0.8 W/m·K, with a temperature difference of 30 K between the two sides.

    A) 24 W
    B) 36 W
    C) 48 W
    D) 60 W

99. Select all the mechanisms involved in convective heat transfer:

    A) Natural convection
    B) Radiation
    C) Natural convection
    D) Forced convection

100.	Fill in the blank: The emissivity of a perfect black body is _____.

101.	Determine the heat transfer rate by radiation from a surface at 500 K with an area of 2 m² and emissivity of 0.9. The Stefan-Boltzmann constant is $5.67 \times 10^{-8}\text{ W/m}^2\text{K}^4$.

102.	Calculate the total heat transferred by a 1.5 m long, 0.01 m diameter cylindrical rod with thermal conductivity of 15 W/m·K, when the temperature difference between its ends is 100 K.

103.	Which of the following instruments is used to measure electrical current?

A) Voltmeter
B) Ammeter
C) Ohmmeter
D) Thermometer

104.	Select all the correct statements about PID controllers:

A) The proportional term helps decrease the rise time.
B) The integral term aids in eliminating the steady-state error.
C) The derivative term assists in reducing the overshoot.
D) The proportional term helps in removing the steady-state error.

105.	Fill in the blank: A _____ is used to measure temperature and can be used in feedback control systems.

106.	Determine the resistance of a thermistor at 25°C if its resistance at 0°C is 1000 ohms and the temperature coefficient is -0.04/°C.

A) 800 ohms
B) 900 ohms
C) 1000 ohms
D) 1100 ohms

107. Select all the instruments that can be used to measure pressure:

A) Barometer
B) Manometer
C) Tachometer
D) Anemometer

108. A differential pressure sensor is used to measure the flow rate of a fluid in a pipe. If the pressure difference is 50 Pa and the density of the fluid is 1000 kg/m³, what is the flow rate in m/s?

A) 0.1 m/s
B) 0.5 m/s
C) 0.316 m/s
D) 2.0 m/s

109. Select all the true statements about strain gauges:

A) They measure strain by changing resistance.
B) They are used in load cells.
C) The gauge factor is the ratio of relative change in electrical resistance to the mechanical strain.
D) They are unaffected by temperature changes.

2 EXAM PREP.

1. Determine the equation of the ellipse that fits through the points (0, 4), (2, 0), and (-1, -3).

   A) $\dfrac{x^2}{4} + \dfrac{y^2}{16} = 1$

   B) $\dfrac{x^2}{16} + \dfrac{y^2}{4} = 1$

   C) $\dfrac{x^2}{9} + \dfrac{y^2}{9} = 1$

   D) $\dfrac{x^2}{25} + \dfrac{y^2}{4} = 1$

2. Select all the correct statements about the angle between vectors:

   A) The dot product of two vectors can be used to find the angle between them.
   B) The angle between two vectors is always between 0° and 180°.
   C) The cross product of two vectors can also determine the angle between them.
   D) The angle between two vectors is independent of their magnitudes.

3. Fill in the blank: The volume of the solid enclosed by the paraboloid $z = x^2 + y^2$ and the plane z=4 is _____.

4. Solve for zzz in the equation $z^2 + (3 + 4i)z + 5i = 0.$

   A) $z = -1 + i$ e $z = -2 - 5i$

   B) $z = -2 + i$ e $z = -1 - 5i$

   C) $z = -2 + 5i$ e $z = -1 - i$

   D) $z = -1 + 5i$ e $z = -2 - i$

5. Calculate the expected value and variance of a discrete random variable X that takes values 1, 2, and 3 with probabilities 0.2, 0.5, and 0.3 respectively.

6. Find the eigenvalues of the matrix $\begin{pmatrix} 3 & 1 \\ 1 & 3 \end{pmatrix}$.

   A) 4,2
   B) 1,5
   C) 3,3
   D) 0,6

7. A die is rolled twice. What is the probability of getting a sum of 7?

   A) $\frac{1}{6}$
   B) $\frac{1}{12}$
   C) $\frac{1}{8}$
   D) $\frac{1}{36}$

8. Select all the correct statements about the normal distribution:

   A) It is symmetric about the mean.
   B) Approximately 95% of the data falls within one standard deviation of the mean.
   C) The mean, median, and mode are all equal.
   D) It has a bell-shaped curve.

9. Fill in the blank: The _____ is the probability distribution of a discrete random variable with only two possible outcomes, typically labeled success and failure.

10. Calculate the variance of the following dataset: {2, 4, 4, 4, 5, 5, 7, 9}.

11. A coin is flipped 10 times. What is the probability of getting exactly 6 heads?

    A) $\binom{10}{6} \left(\frac{1}{2}\right)^{10}$
    B) $\binom{10}{6} \left(\frac{1}{2}\right)^{6} \left(\frac{1}{2}\right)^{4}$
    C) $\binom{10}{6} \left(\frac{1}{2}\right)^{4} \left(\frac{1}{2}\right)^{6}$
    D) All of the above

12. According to the NSPE Code of Ethics, engineers should hold paramount the safety, health, and welfare of:

    A) Themselves
    B) The public
    C) Their clients
    D) Their employers

13. Select all the correct actions an engineer should take if they find a serious design flaw in a project:

    A) Ignore it to avoid delays
    B) Report it to the appropriate authority
    C) Correct the flaw quietly without informing anyone
    D) Document the flaw and the steps taken to address it

14. Fill in the blank: Engineers must perform services only in areas of their competence and should undertake assignments only when qualified by _____.

15. An engineer is offered a substantial gift by a contractor during a bidding process. According to ethical guidelines, the engineer should:

    A) Accept the gift to maintain good relations
    B) Refuse the gift to avoid a conflict of interest
    C) Accept the gift but disclose it to their employer
    D) Refuse the gift and report the offer to their professional association

16. Select all the scenarios that would constitute a conflict of interest for an engineer: A) Owning stock in a competitor's company
    B) Accepting a paid consultancy with a vendor while working on a related project
    C) Volunteering for a non-profit organization
    D) Having a close relative working for a client's organization

17. A project requires an initial investment of $50,000 and is expected to generate annual cash flows of $10,000 for 7 years. What is the payback period?

    A) 4 years
    B) 5 years
    C) 6 years
    D) 7 years

18. Select all the true statements about Net Present Value (NPV):

    A) NPV accounts for the time value of money.
    B) A positive NPV indicates a profitable project.
    C) NPV is unaffected by the discount rate.
    D) NPV can be used to compare different projects.

19. Fill in the blank: The _____ rate is the discount rate that makes the Net Present Value (NPV) of a project zero.

20. Calculate the Future Value (FV) of $5,000 invested for 5 years at an annual interest rate of 6%.

A) $6,691.13
B) $6,500.00
C) $6,733.38
D) $6,755.68

21. Select all the factors that affect the internal rate of return (IRR) of a project:

    A) Initial investment cost
    B) Duration of the project
    C) Annual cash flows
    D) Inflation rate

22. A resistor of 10 ohms is connected in series with a 20-ohm resistor and a 12V battery. What is the current flowing through the circuit?

    A) 0.4 A
    B) 0.6 A
    C) 0.8 A
    D) 1.0 A

23. Select all the correct statements about magnetic fields:

    A) Magnetic field lines emerge from the north pole and enter the south pole.
    B) Magnetic field strength is measured in Tesla.
    C) A changing magnetic field can induce an electric current.
    D) Magnetic fields are always created by permanent magnets.

24. Fill in the blank: The _____ law states that the induced electromotive force in any closed circuit is equal to the negative of the time rate of change of the magnetic flux through the circuit.

25. Calculate the power dissipated in a 5-ohm resistor with a current of 3 A flowing through it.

    A) 15 W
    B) 30 W
    C) 45 W
    D) 60 W

26. Select all the correct statements about capacitors:

    A) Capacitors store energy in the form of an electric field.
    B) The capacitance of a capacitor is directly proportional to the surface area of the plates.
    C) The unit of capacitance is the Farad.
    D) Capacitors resist changes in current.

27. Calculate the reactions at the supports A and B for a simply supported beam carrying a uniform load of 300 N/m over its entire 6 m length.

    A) $A = 900\,\text{N},\ B = 900\,\text{N}$

    B) $A = 1800\,\text{N},\ B = 1800\,\text{N}$

    C) $A = 600\,\text{N},\ B = 600\,\text{N}$

    D) $A = 300\,\text{N},\ B = 300\,\text{N}$

28. Select all the true statements about cable tension:

    A) The tension in the vertical cable equals the weight of the mass.
    B) The tension in the inclined cable is greater than the tension in the vertical cable.
    C) Both cables experience the same tension.
    D) The sum of the vertical components of the tensions equals the weight of the mass.

29. Fill in the blank: The centroid of a T-shaped figure formed by three rectangles can be found by taking the _____ of the centroids of the individual rectangles.

30. Calculate the moment of inertia for a composite area consisting of a rectangle (200 mm x 400 mm) and a semicircle (radius 100 mm) about an axis at the base of the rectangle.

    A) $1.67 \times 10^7\,\text{mm}^4$

    B) $3.33 \times 10^7\,\text{mm}^4$

    C) $5.00 \times 10^7\,\text{mm}^4$

    D) $6.67 \times 10^7\,\text{mm}^4$

31. Calculate the minimum coefficient of friction required to prevent slipping for a 200 kg block resting on a 30-degree inclined plane.

    A) 0.5
    B) 0.6
    C) 0.7
    D) 0.8

32. Select all the zero-force members in a simple truss consisting of five members arranged in a triangle with an extension on one side:

    A) Top chord member
    B) Bottom chord member
    C) Diagonal member
    D) Vertical member

33. Determine the maximum bending moment in a beam fixed at both ends and subjected to a central point load of 1000 N.

    A) 1250 Nm
    B) 2500 Nm
    C) 3750 Nm
    D) 5000 Nm

34. Fill in the blank: A column's stability under a compressive load depends on its _____ and _____.

35. Draw the shear and moment diagrams for a simply supported beam carrying a point load of 500 N at the midpoint and a uniformly distributed load of 200 N/m over its 4-meter length.

36. Calculate the hydrostatic pressure at the bottom of a water tank that is 3 meters deep.

    A) 10 kPa
    B) 20 kPa
    C) 30 kPa
    D) 40 kPa

37. Calculate the velocity of a car initially at rest after it accelerates at 2.5 m/s² for 12 seconds.

    A) 15 m/s
    B) 25 m/s
    C) 30 m/s
    D) 35 m/s

38. Select all the true statements about tension in multiple cables:

    A) The tension in the vertical cable equals the weight of the mass.
    B) The tension in the cable at 60° is greater than the tension in the cable at 45°.
    C) Both cables experience the same tension.
    D) The sum of the vertical components of the tensions equals the weight of the mass.

39. Fill in the blank: The coefficient of kinetic friction required to keep a 200 kg block from accelerating down a 25° incline is _____.

40. Determine the frequency of a mass-spring system with a mass of 1.5 kg and a spring constant of 300 N/m.

    A) 1.41 Hz
    B) 2.21 Hz
    C) 3.54 Hz
    D) 4.37 Hz

41. Calculate the impulse needed to change the velocity of a 500 kg vehicle from 15 m/s to 5 m/s.

    A) 2000 Ns
    B) 3000 Ns
    C) 4000 Ns
    D) 5000 Ns

42. Select all the correct statements about work done by a force:

    A) Work is the product of force and displacement.
    B) Work done is zero if the force is perpendicular to the displacement.
    C) Work done is maximum when the force is parallel to the displacement.
    D) Work is a scalar quantity.

43. Fill in the blank: The period of a pendulum with a length of 1.5 meters is _____.

44. Calculate the centripetal force acting on a 1 kg object moving at 10 m/s in a circle of radius 2 meters.

45. Determine the range of a projectile launched at an initial speed of 20 m/s at an angle of 30°.

46. Find the frequency of vibration for a beam clamped at both ends, with a length of 4 meters, a mass per unit length of 3 kg/m, and a flexural rigidity of 2000 Nm².

47. Calculate the hoop stress in a hollow cylindrical pressure vessel with an internal radius of 500 mm, a wall thickness of 10 mm, and an internal pressure of 5 MPa.

    A) 25 MPa
    B) 50 MPa
    C) 75 MPa
    D) 100 MPa

48. Select all the correct statements about elastic modulus:

    A) It is a measure of a material's stiffness.
    B) It is calculated as the ratio of stress to strain.
    C) It is the same for all materials.
    D) Higher elastic modulus indicates a stiffer material.

49. Fill in the blank: The bending stress at the bottom fiber of a rectangular beam subjected to a bending moment can be calculated using the formula _____.

50. Determine the energy stored in a spring with a stiffness of 150 N/m compressed by 40 mm.

    A) 0.12 J
    B) 0.24 J
    C) 0.36 J
    D) 0.48 J

51. Calculate the shear stress in a bolt that transmits a force of 8000 N across two plates. The bolt diameter is 12 mm.

52. Select all the effects of Poisson's ratio:

    A) It describes the ratio of lateral strain to longitudinal strain.
    B) It is always greater than 1.
    C) It can be used to determine lateral strain when longitudinal stress is known.
    D) It is dimensionless.

53. Fill in the blank: The critical load for buckling in a pinned-pinned column can be determined using _____ formula.

54. Calculate the stress developed in a steel bar fixed at both ends when the temperature increases by 50°C. Assume the coefficient of thermal expansion is 12×10^{-6} /°C and the modulus of elasticity is 210 GPa.

55. Determine the angle of twist per meter length for a steel shaft subjected to a torsional moment of 360 Nm. The shaft has a diameter of 50 mm and a shear modulus of 80 GPa.

56. Calculate the maximum shear stress in a wide-flange beam subjected to a vertical shear force of 30 kN. The beam's cross-sectional area is 4000 mm².

    A) 3.75 MPa
    B) 5.00 MPa
    C) 7.50 MPa
    D) 10.00 MPa

57. Which of the following materials has the highest thermal conductivity?

    A) Aluminum
    B) Copper
    C) Steel
    D) Titanium

58. Select all the true statements about heat treatment processes:

    A) Annealing improves ductility.
    B) Quenching increases hardness.
    C) Tempering reduces brittleness.
    D) Normalizing creates a uniform grain structure.

59. Fill in the blank: The hardness of a material can be measured using the _____ scale.

60. Calculate the density of a material if a sample has a mass of 50 grams and a volume of 20 cm³.

    A) 1.5 g/cm³
    B) 2.0 g/cm³
    C) 2.5 g/cm³
    D) 3.0 g/cm³

61. Select all the correct properties of ceramics:

    A) High melting point
    B) Low thermal conductivity
    C) High brittleness
    D) High electrical conductivity

62. Fill in the blank: The process of adding carbon to the surface of steel to improve its hardness is called _____.

63. Which of the following materials is most suitable for high-temperature applications due to its high melting point?

    A) Aluminum
    B) Brass
    C) Tungsten
    D) Zinc

64. Select all the true statements about composite materials:

    A) They are made from two or more constituent materials.
    B) They typically have improved mechanical properties.
    C) They are always homogeneous.
    D) Fiberglass is an example of a composite material.

65. Calculate the hydraulic gradient required to transport water through a 100 m long pipe with a diameter of 200 mm, assuming a flow rate of 0.05 m³/s and a friction factor of 0.02.

    A) 0.005
    B) 0.013
    C) 0.015
    D) 0.02

66. Select all the correct statements about orifice flow rate:

    A) The flow rate increases with the square root of the pressure difference.
    B) The flow rate is directly proportional to the orifice diameter.
    C) The discharge coefficient affects the flow rate.
    D) The flow rate is inversely proportional to the fluid viscosity.

67. Fill in the blank: The effect of viscosity on the velocity profile in a pipe can be described using the _____ equation.

68. Calculate the power required by a pump to lift water to a height of 30 m, assuming a flow rate of 0.1 m³/s and pump efficiency of 70%.

    A) 2.86 kW
    B) 21.02 kW
    C) 6.43 kW
    D) 8.57 kW

69. Determine the capillary rise in a tube with a diameter of 1.5 mm when immersed in water, given the surface tension of water as 0.072 N/m and the contact angle as zero.

70. Select all the correct statements about flow over weirs:

    A) The flow rate over a rectangular weir is proportional to the square root of the head.
    B) The width of the weir affects the flow rate.
    C) The discharge coefficient is used to account for energy losses.
    D) The flow rate over a weir is independent of the weir shape.

71. Fill in the blank: The pressure surge due to water hammer can be estimated using the _____ equation.

72. Calculate the settling velocity of a particle in water, given the particle diameter of 0.5 mm and specific gravity of 2.5.

    A) 0.045 m/s
    B) 0.204 m/s
    C) 0.135 m/s
    D) 0.180 m/s

73. Determine the lift force generated by a hydrofoil moving at 15 m/s, with a chord length of 1 m and a lift coefficient of 0.8.

74. Assess whether the flow is laminar or turbulent in a pipe with a diameter of 25 mm, given a Reynolds number of 4000.

    A) Laminar
    B) Turbulent
    C) Transition
    D) None of the above

75. A gas undergoes an isothermal expansion at 300 K from an initial volume of 1 m³ to a final volume of 3 m³. What is the change in internal energy of the gas?

    A) 0 J
    B) 500 J
    C) 1000 J
    D) 1500 J

76. Select all the true statements about the first law of thermodynamics:

    A) Energy can be created or destroyed.
    B) The change in internal energy is equal to the heat added to the system minus the work done by the system.
    C) It applies to closed systems only.
    D) It is a statement of the conservation of energy.

77. Fill in the blank: The _____ cycle is an idealized thermodynamic cycle that describes the functioning of a heat engine and is used to determine the maximum possible efficiency.

78. Calculate the efficiency of a Carnot engine operating between temperatures of 600 K and 300 K.

    A) 25%
    B) 50%
    C) 75%
    D) 100%

79. Determine the specific heat capacity of a substance if 2000 J of heat is required to raise the temperature of 2 kg of the substance by 5°C.

80. Select all the true statements about entropy:

    A) Entropy measures the disorder of a system.
    B) Entropy always decreases in a spontaneous process.
    C) The entropy of a perfect crystal at absolute zero is zero.
    D) Entropy is a state function.

81. Fill in the blank: The _____ law of thermodynamics states that the total entropy of an isolated system can never decrease over time.

82. A heat pump extracts 2000 J of heat from a cold reservoir and releases 2500 J of heat to a hot reservoir. Calculate the work done by the heat pump.

    A) 300 J
    B) 400 J
    C) 500 J
    D) 600 J

83. Determine the final temperature when 500 g of water at 80°C is mixed with 500 g of water at 20°C.

84. Calculate the change in enthalpy for a process where the internal energy change is 500 J, and the work done by the system is 200 J.

    A) 300 J
    B) 500 J
    C) 700 J
    D) 900 J

85. A wall is insulated with a material that has a thermal conductivity of 0.04 W/m·K. If the wall is 0.1 m thick and the temperature difference across it is 20°C, what is the heat flux through the wall?

    A) 8 W/m²
    B) 10 W/m²
    C) 20 W/m²
    D) 40 W/m²

86. Select all the true statements about convection heat transfer:

    A) Natural convection occurs due to temperature-induced density differences.
    B) Forced convection involves fluid motion generated by external means like fans or pumps.
    C) The convective heat transfer coefficient is independent of fluid velocity.
    D) Convection only occurs in liquids.

87. Fill in the blank: The rate of heat transfer by radiation between two surfaces is described by the _____ law.

88. Determine the amount of heat transferred through a 0.5 m² area of a flat surface with an emissivity of 0.85 and a temperature difference of 40°C over 2 hours.

89. Calculate the heat transfer rate by conduction through a cylindrical pipe with an inner radius of 0.05 m, outer radius of 0.1 m, length of 1 m, thermal conductivity of 0.5 W/m·K, and a temperature difference of 100°C.

    A) 453.24 W
    B) 62.8 W
    C) 94.2 W
    D) 125.6 W

90. Select all the correct statements about heat exchangers:

    A) They transfer heat between two or more fluids.
    B) They are used only in heating applications.
    C) The effectiveness of a heat exchanger is the ratio of actual heat transfer to the maximum possible heat transfer.
    D) They can be classified as direct or indirect contact heat exchangers.

91. Fill in the blank: The _____ number is a dimensionless number used in convective heat transfer to characterize the relative importance of conduction and convection.

92. A heat pump has a coefficient of performance (COP) of 3.5 and absorbs 5000 J of heat from the cold reservoir. How much work does the heat pump require to operate?

    A) 1428.6 J
    B) 2000 J
    C) 2500 J
    D) 3500 J

93. A thermocouple is used to measure temperature and produces a voltage of 25 mV at 100°C. If the thermocouple has a sensitivity of 0.25 mV/°C, what is the ambient temperature when it reads 10 mV?

    A) 20°C
    B) 30°C
    C) 40°C
    D) 50°C

94. Select all the true statements about PID controllers:

    A) The proportional term determines the reaction to the current error.
    B) The integral term determines the reaction to the rate of change of error.
    C) The derivative term predicts future error based on the rate of change.
    D) PID controllers are used only in temperature control systems.

95. Fill in the blank: The _____ is a measure of the accuracy of an instrument, defined as the maximum deviation observed from the true value.

96. Calculate the signal-to-noise ratio (SNR) in dB if the signal power is 20 mW and the noise power is 0.5 mW.

    A) 13 dB
    B) 16 dB
    C) 20 dB
    D) 26 dB

97. Select all the correct statements about strain gauges:

    A) They measure deformation or strain in a material.
    B) The resistance of a strain gauge changes with applied force.
    C) They are only used in static measurements.
    D) They are commonly used in load cells and pressure sensors.

98. Fill in the blank: The _____ law relates the induced voltage in a coil to the rate of change of magnetic flux through the coil.

99. A cantilever beam is loaded with a uniform load of 500 N/m over its length of 3 meters. What is the maximum bending moment at the fixed end?

    A) 225 Nm
    B) 450 Nm
    C) 675 Nm
    D) 900 Nm

100. Select all the true statements about fatigue failure:

    A) Fatigue failure occurs under cyclic loading.
    B) The endurance limit is the stress level below which fatigue failure will not occur.
    C) Fatigue life decreases with increasing mean stress.
    D) Surface finish has no effect on fatigue strength.

101. Fill in the blank: The _____ factor is used to account for uncertainties in the design process to ensure safety and reliability.

102. Calculate the deflection at the midpoint of a simply supported beam of length 4 meters, subjected to a central point load of 800 N. The flexural rigidity (EI) of the beam is 2000 Nm^2.

    A) 0.533 m
    B) 0.008 m
    C) 0.012 m
    D) 0.016 m

103. Select all the correct statements about gears:

    A) Spur gears have teeth parallel to the axis of rotation.
    B) Helical gears have teeth that are cut at an angle to the face of the gear.
    C) Bevel gears are used to transmit motion between intersecting shafts.
    D) Worm gears can achieve high gear ratios in a single stage.

104. Fill in the blank: The _____ criterion is used to predict the onset of yielding in ductile materials under complex loading conditions.

105. A steel shaft transmits a power of 20 kW at 1500 rpm. Calculate the torque transmitted by the shaft.

106. Select all the true statements about bolted joints:

    A) Preloading a bolt increases the joint's fatigue strength.
    B) The clamping force in a bolted joint is independent of the applied torque.
    C) Bolted joints are typically designed to prevent slippage under load.
    D) Using washers can help distribute the load and reduce the risk of loosening.

107. Fill in the blank: The _____ ratio is the ratio of the ultimate tensile strength to the yield strength of a material.

108. A cylindrical pressure vessel with a radius of 0.5 meters and a wall thickness of 10 mm is subjected to an internal pressure of 2 MPa. Calculate the hoop stress.

    A) 100 MPa
    B) 200 MPa
    C) 300 MPa
    D) 400 MPa

109. Select all the correct statements about bearings:

    A) Ball bearings are suitable for high-speed applications.
    B) Roller bearings can carry higher radial loads than ball bearings.
    C) Thrust bearings are designed to accommodate axial loads.
    D) Plain bearings are also known as journal bearings.

110. Fill in the blank: The _____ analysis method is used to determine the stresses and deformations in a structure by dividing it into smaller, simpler parts called elements.

3 EXAM PREP.

1. Find the function y(x) that extremizes the functional
$$J[y] = \int_0^1 (y'^2 - y^2 + 6y)\, dx$$
with boundary conditions y(0)=0 and y(1)=0

2. Evaluate the integral of the function $f(z) = \frac{z+1}{z^2+1}$ around a circle centered at z=0 with radius 2.

    A) 2πi
    B) πi
    C) 0
    D) −πi

3. Solve the system of linear equations using the Gauss-Seidel method up to two iterations, starting from the initial guess (0,0,0):
$$10x + 2y - z = 3$$
$$3x + 10y + z = 15$$
$$-x + y + 10z = 27$$

4. Determine the number of ways to arrange 4 red, 3 green, and 3 blue balls such that no two blue balls are adjacent.

    A) 1260
    B) 2520
    C) 5040
    D) 10080

5. Find the partial derivatives of the function $f(x, y) = x^2 y + e^{xy} + y^3$ with respect to x and y at the point (1,2).

6. For a continuous random variable X with probability density function $f(x) = \frac{1}{2} e^{-|x|}$, calculate the expected value and variance.

7. The probability of getting exactly two heads in three flips of a fair coin is:

    A) 1/2
    B) 1/4
    C) 1/8
    D) 3/8

8. Select all the true statements about the normal distribution:

   A) It is symmetric about the mean.
   B) The mean, median, and mode are all equal.
   C) It has a bell-shaped curve.
   D) The total area under the curve is not equal to 1.

9. Fill in the blank: The _____ theorem states that, for a sufficiently large sample size, the sampling distribution of the sample mean will be normally distributed regardless of the population distribution.

10. Given a dataset: 4, 8, 6, 5, 3, calculate the median.

11. A box contains 3 red, 2 green, and 5 blue balls. Two balls are drawn at random without replacement. What is the probability that both balls are red?

    A) 1/10
    B) 1/15
    C) 1/20
    D) 1/45

12. An engineer discovers a critical safety issue in a product that has already been released to the market. According to the NSPE Code of Ethics, what is the engineer's obligation?

    A) Ignore the issue to avoid liability.
    B) Report the issue to the appropriate authorities and stakeholders.
    C) Try to fix the issue without informing anyone.
    D) Keep the information confidential and not report it.

13. Select all the true statements about conflicts of interest in professional practice:

    A) A conflict of interest can occur if an engineer has a financial interest in a project they are working on.
    B) Engineers must disclose any potential conflicts of interest to their employer or client.
    C) Accepting gifts from suppliers always constitutes a conflict of interest.
    D) Engineers can ignore minor conflicts of interest if they believe it will not impact their work.

14. Fill in the blank: The _____ principle in engineering ethics requires that engineers hold paramount the safety, health, and welfare of the public in the performance of their professional duties.

15. If an engineer makes a professional mistake that causes harm, what should they do according to the ethical guidelines?

    A) Hide the mistake to protect their reputation.
    B) Acknowledge the mistake and take responsibility.
    C) Blame others for the mistake.
    D) Leave the job to avoid consequences.

16. Select all the responsibilities engineers have towards their clients:

    A) Provide services only in areas of their competence.
    B) Keep client information confidential unless required by law to disclose it.

C) Accept assignments outside their field of expertise if it benefits the client.
D) Act as faithful agents or trustees for their clients.

17. If an investment of $10,000 earns an annual interest rate of 5%, compounded annually, what will be its value at the end of 5 years?

    A) $12,500
    B) $12,762.82
    C) $13,276.25
    D) $13,810.00

18. Select all the true statements about Net Present Value (NPV):

    A) NPV accounts for the time value of money.
    B) A positive NPV indicates a profitable investment.
    C) NPV is the same as the internal rate of return (IRR).
    D) NPV can be used to compare different investment options.

19. Fill in the blank: The _____ is the interest rate at which the net present value of all cash flows from a particular project is zero.

20. A machine costs $20,000 and is expected to generate annual savings of $5,000 for 5 years. If the company's discount rate is 8%, what is the Present Value (PV) of the savings?

    A) $19,000
    B) $20,000
    C) $19,964
    D) $23,150

21. Select all the correct statements about depreciation:

    A) Straight-line depreciation spreads the cost evenly over the asset's useful life.
    B) Declining balance depreciation results in higher expenses in the earlier years.
    C) Depreciation affects the book value of an asset but not the cash flow.
    D) Salvage value is not considered in calculating depreciation.

22. Calculate the resistance of a wire with a length of 2 meters, a cross-sectional area of 1 mm², and a resistivity of 1.68×10^{-8} Ω·m.

   A) 3.36×10^{-5} Ω
   B) 3.36×10^{-3} Ω
   C) 1.68×10^{-5} Ω
   D) 1.68×10^{-3} Ω

23. Select all the true statements about capacitors in AC circuits:

   A) The capacitive reactance decreases with increasing frequency.
   B) The current leads the voltage by 90 degrees.
   C) Capacitors store energy in the form of an electric field.
   D) In a purely capacitive circuit, the power factor is zero.

24. Fill in the blank: The _____ law states that the line integral of the magnetic field around a closed loop is equal to the permeability times the electric current passing through the loop.

25. A transformer has 500 turns on the primary coil and 100 turns on the secondary coil. If the primary voltage is 240 V, what is the secondary voltage?

   A) 48 V
   B) 60 V
   C) 120 V
   D) 480 V

26. Select all the correct statements about electromagnetic waves:

   A) They can travel through a vacuum.
   B) The electric and magnetic fields are perpendicular to each other and to the direction of propagation.
   C) The speed of electromagnetic waves in a vacuum is 3×10^8 m/s.
   D) They require a medium to propagate.

27. Determine the necessary force at point B to keep a beam in horizontal equilibrium which is hinged at one end (point A) and has a point load of 500 N at 2 meters from point A.

   A) 1000 N
   B) 750 N
   C) 500 N
   D) 250 N

28. Find the centroid of a T-shaped object formed by two rectangles, one 150 mm by 300 mm and the other 150 mm by 150 mm, with the smaller rectangle centered on top of the larger one.

29. Calculate the moment of inertia for a right triangle with base 200 mm and height 400 mm about its base.

   A) $1.07 \times 10^6 \text{ mm}^4$

   B) $2.67 \times 10^6 \text{ mm}^4$

   C) $3.17 \times 10^6 \text{ mm}^4$

   D) $4.27 \times 10^6 \text{ mm}^4$

30. Determine the maximum load that can be placed on a 25° inclined plane without slipping, given the coefficient of static friction is 0.35.

31. Find the tension in two cables holding a 600 kg mass in equilibrium, where one cable makes a 30° angle with the vertical and the other 60°.

32. Calculate the shear force at a point 3 m from the left support in a simply supported beam 9 m long with a point load of 800 N at the center.

   A) 133.33 N
   B) 266.67 N
   C) 400.00 N
   D) 533.33 N

33. Determine the bending moment at the midpoint of a simply supported beam 5 m long with a uniformly distributed load of 400 N/m.

34. Assess the stability of a 4 m tall column fixed at one end and free at the other. The column has a cross-sectional area of 100 mm^2 and a moment of inertia I=8.33×10−6 m^4. If the column is made of steel with a Young's Modulus E=200 GPa, determine the critical load that will cause the column to buckle.

35. Calculate the forces in all members of a simple truss consisting of four members forming a square, with a load of 1200 N applied at the top node. The truss is symmetric, and each member has a length of 3 meters.

36. Determine the force required to maintain equilibrium in a pulley system with a mass of 250 kg hanging from one side.

   A) 2450 N
   B) 2500 N
   C) 2550 N
   D) 2600 N

37. Calculate the final velocity of an object sliding down a 45° inclined plane, starting from rest with a constant acceleration of 5 m/s² over a distance of 150m.

    A) 35 m/s
    B) 38.7 m/s
    C) 42.4 m/s
    D) 45.0 m/s

38. Calculate the tensions in three cables supporting a 150 kg mass in equilibrium, where the cables make angles of 30°, 45°, and 60° with the vertical.

39. Determine the minimum coefficient of static friction required to prevent a 100 kg block from sliding down an incline that increases its angle from 10° to 30°.

    A) 0.176
    B) 0.225
    C) 0.577
    D) 0.5

40. Determine the damped natural frequency of a mass-spring system with a mass of 1 kg, a spring constant of 200 N/m, and a damping coefficient of 5 Ns/m.

41. Calculate the change in momentum for a 1000 kg vehicle that decelerates from 100 m/s to 70 m/s over 10 seconds.

    A) 30,000 kg·m/s
    B) 35,000 kg·m/s
    C) 40,000 kg·m/s
    D) 45,000 kg·m/s

42. Determine the kinetic energy lost by a 200 kg cart colliding elastically with a stationary 300 kg cart. Answer:

43. Find the time it takes for a pendulum with a length of 2 meters to complete one full swing (back and forth).

    A) 2.83 s
    B) 3.01 s
    C) 3.19 s
    D) 3.28 s

44. Calculate the necessary centripetal force to keep a 2 kg object moving in a circle of radius 1 meter at a constant speed of 15 m/s.

    A) 300 N
    B) 450 N
    C) 500 N
    D) 600 N

45. Determine the initial launch speed required for a projectile to reach a maximum height of 50 meters when launched at an angle of 60°.

46. Calculate the amplitude of vibration for a beam clamped at one end, free at the other, excited at its natural frequency. Assume the beam's total length is 5 meters, mass per unit length is 1 kg/m, and has a flexural rigidity of 500 Nm^2.

47. Determine the normal stress in a welded joint that must support a load of 1500 N, if the cross-sectional area of the weld is 25 mm².

    A) 60 MPa

    B) 80 MPa

    C) 40 MPa

    D) 50 MPa

48. Calculate the maximum deflection in the center of a simply supported beam 4 meters long, carrying a point load of 500 N at its midpoint. Assume E (Young's modulus) = 200 GPa and I (moment of inertia) = 5000 cm⁴.

    A) 0.0125 mm

    B) 0.025 mm

    C) 0.0667 mm

    D) 0.100 mm

49. A rod is subjected to a tensile force of 2000 N and a torsional moment of 50 Nm. If the rod's diameter is 15 mm, calculate the equivalent von Mises stress.

    A) 55.5 MPa

    B) 75.3 MPa

    C) C) 131.17 MPa

    D) 95.1 MPa

50. Determine the shear force at a notch in a beam that carries a uniformly distributed load of 1000 N/m over a span of 2 meters.

    A) 500 N

    B) 1000 N

    C) 1500 N

    D) 2000 N

51. Calculate the elastic buckling stress for a column with an effective length of 3 meters, a cross-sectional area of 2000 mm², and a modulus of elasticity of 210 GPa.

    A) 78.5 MPa

1. B) 5757.27 MPa

    C) 95.0 MPa

    D) 102.5 MPa

52. Determine the moment of inertia for a T-shaped cross-section with the following dimensions:
    - Flange width: 200 mm
    - Flange thickness: 10 mm
    - Web height (total height including flange): 100 mm
    - Web thickness: 20 mm

    A) 1.67×10^6 mm^4

    B) 2.33×10^6 mm^4

    C) 3.00×10^6 mm^4

    D) 4.25×10^6 mm^4

53. Estimate the critical speed of a rotating shaft with a diameter of 25 mm and a length of 2 meters, considering it as a simply supported beam.

    A) 50 rad/s

    B) 5654.7 rad/s

    C) 100 rad/s

    D) 125 rad/s

54. Given a longitudinal strain of 0.002 and a lateral strain of -0.0005 in a material, calculate the Poisson's ratio.

55. Determine the maximum normal stress at a point in a stressed body where the existing stresses are σx = 60 MPa, σy = 40 MPa, and τxy = 20 MPa.

    A) 72.36 MPa.

    B) 80 MPa

    C) 90 MPa

D) 100 MPa

56. Calculate the energy absorbed by a cylindrical bumper made of steel that deforms 5 cm under a load of 10000 N. Assume the bumper acts like a spring.

    A) 125 J

    B) 250 J

    C) 375 J

    D) 500 J

57. **Tensile Stress in a Rod** Calculate the tensile stress in a steel rod that is 10 mm in diameter and is subjected to a force of 1,000 N.

    A) 12.73 MPa

    B) 127.3 MPa

    C) 1,273 MPa

    D) 12,730 MPa

58. **Strain in a Steel Bar** Given a steel bar under a tensile stress of 100 MPa and a Young's modulus of 210 GPa, the strain experienced by the bar is _____.

59. **Deflection of a Cantilever Beam** Consider a 2-meter long cantilever beam with a point load of 500 N at the free end. Given a Young's modulus of 200 GPa and a moment of inertia of 4000 cm^4, select all possible deflections at the free end:

    A) 0.025 mm

    B) 0.25 mm

    C) 2.5 mm

    D) 25 mm

60. **Shear Stress in a Circular Shaft** What is the shear stress in a circular shaft 25 mm in diameter transmitting a torque of 300 Nm?

    A) 4.56 MPa

    B) 45.6 MPa

    C) 1527.89 MPa

    D) 4560 MPa

61. **Compressive Stress in a Column** The compressive stress in a column with a 50 cm^2 cross-sectional area, subjected to a 10,000 N axial load, is _____ MPa.

62. **Thermal Expansion in a Rod** Determine the change in length of a 2-meter long aluminum rod when heated by 80°C. The coefficient of thermal expansion for aluminum is 23×10−6/°C23 ₩times 10^{−6} /°C23×10−6/°C.

    A) 0.36 mm

    B) 3.68 mm

    C) 36.8 mm

    D) 368 mm

63. **Bending Stress in an I-Beam** An I-beam subjected to a bending moment of 10 kNm has a moment of inertia of 8000 cm⁴ and is 150 mm from the neutral axis to the extreme fibers. Which of the following could be the bending stress at the extreme fibers?

    A) 18.75 MPa

    B) 187.5 MPa

    C) 1875 MPa

    D) 18750 MPa

64. **Buckling Load of a Slender Column** A slender steel column, 3 meters in length with a radius of gyration of 50 mm and a modulus of elasticity of 200 GPa, has a critical buckling load of _____ kN.

65. **Stress Concentration Around a Hole** Calculate the maximum stress concentration around a 20 mm diameter hole in a plate under a 150 MPa tensile stress. The width of the plate is 100 mm.

    A) 165 MPa

    B) 175 MPa

    C) 185 MPa

    D) 450 MPa.

66. **Impact Force in a Falling Rod** A steel rod 1 meter in length and 5 kg in mass falls from a height of 2 meters. If the contact time upon impact is 0.005 seconds, what are the possible values for the impact force?

    A) 1,000 N

    B) 2,000 N

    C) 6264.18 N.

    D) 20,000 N

67. The pressure gradient in a 200-meter-long pipe with a diameter of 0.1 meter, when the flow rate is 1 cubic meter per second and the fluid viscosity is 0.8 millipascal-second, is _____ Pa/m.

68. Using a Venturi meter, what is the flow rate if the differential pressure is 250 Pa and the throat to pipe diameter ratio is 0.5?

    - A) 0.5 m³/s
    - B) 0.00136 m³/s
    - C) 0.9 m³/s
    - D) 1.1 m³/s

69. Calculate the critical depth in a rectangular channel with a flow rate of 10 m³/s and a width of 5 meters.

    - A) 8.0 m/s^2
    - B) 9.0 m/s^2
    - C) 9.81m/s^2
    - D) 10.1 m/s^2

70. The downstream depth of a hydraulic jump in a 2-meter-wide channel, where the upstream depth is 0.5 meter and the flow rate is 20 cubic meters per second, is _____ meters.

71. What is the discharge coefficient of an orifice plate with a diameter of 0.2 meters, a flow rate of 0.5 cubic meters per second, and a pressure drop of 150 kPa?

    - A) 0.6
    - B) 0.65
    - C) 0.7
    - D) 0.75

72. Identify the correct statements about the force exerted by a fluid accelerating from 0 to 10 m/s in a pipe with a diameter of 0.15 m and a length of 50 m. (Choose all that apply)

    - A) The force can be calculated using the momentum change formula.
    - B) The force is affected by the fluid's density.
    - C) The force depends on the fluid's viscosity.
    - D) The force is 7500 N.

73. The drag force on a sphere with a diameter of 0.1 meter moving at 5 m/s in a fluid with a density of 1000 kg/m³ and a dynamic viscosity of 0.001 Pascal-second is _____ Newtons.

74. What is the sedimentation rate of particles with a diameter of 0.002 meters and a specific gravity of 2.6 in a fluid with a viscosity of 0.001 Pa·s?

    - A) 0.0012 m/s
    - B) 0.0023 m/s
    - C) 0.0034 m/s
    - D) 0.0045 m/s

75. When calculating the power required to operate a pump that lifts 0.3 cubic meters per second of water to a height of 10 meters with an efficiency of 65%, which of the following factors must be considered? (Choose all that apply)

    - A) The gravitational constant
    - B) The density of water
    - C) The viscosity of water
    - D) The mechanical losses in the pump

76. Determine the flow velocity in a conduit with a hydraulic diameter of 0.5 meters, given a flow rate of 2 cubic meters per second.

    - A) 5.1 m/s
    - B) 6.4 m/s
    - C) 8.2 m/s
    - D) 9.5 m/s

77. The entropy change for a reversible process in which 500 J of heat is absorbed by a system at a constant temperature of 250 K is _____ J/K.

78. What is the efficiency of a Carnot engine operating between two heat reservoirs at temperatures of 400 K and 300 K?

- A) 20%
- B) 25%
- C) 33%
- D) 50%

79. Select all true statements about the first law of thermodynamics for a closed system undergoing a cyclic process.

- A) The internal energy change over one cycle is zero.
- B) Heat transfer must equal the work done.
- C) The system returns to its original state at the end of the cycle.
- D) Energy is neither created nor destroyed.

80. The heat transfer required to increase the temperature of 10 kg of water from 20°C to 50°C, assuming no phase change and a specific heat capacity of 4.18 kJ/kg·K, is _____ kJ.

81. A refrigeration cycle removes 500 kJ of heat from a low-temperature reservoir and expels 600 kJ to a high-temperature reservoir. What is the coefficient of performance (COP) of the refrigerator?

- A) 0.83
- B) 1.2
- C) 1.5
- D) 2.0

82. Which of the following are typical applications of the second law of thermodynamics? (Choose all that apply)

- A) Determining the maximum efficiency of heat engines.
- B) Calculating the minimum energy required for a compressor.
- C) Predicting the direction of heat transfer.
- D) Evaluating the performance of refrigeration cycles.

83. The standard molar entropy of a certain gas increases by _____ J/mol·K when it is heated from 25°C to 100°C at constant pressure, given its molar heat capacity at constant pressure is 29 J/mol·K.

84. During a process, a gas does 150 J of work against its surroundings and receives 100 J of heat. What is the change in internal energy of the gas?

- A) -50 J
- B) 50 J
- C) 250 J
- D) -250 J

85. When evaluating a steam power cycle, which parameters are crucial for increasing the thermal efficiency? (Choose all that apply)

- A) Maximizing the boiler pressure.
- B) Minimizing the condenser pressure.
- C) Increasing the superheat temperature.
- D) Reducing the turbine inlet temperature.

86. The power output of a heat engine that absorbs 2000 J of heat and expels 1500 J of waste heat per cycle is _____ J.

87. **Identify the critical components of a feedback control system.**

- A) Sensors
- B) Actuators
- C) Set point
- D) Data logging system

88. **The sensitivity of a sensor is given as 0.01 V/°C. If the temperature changes by 25°C, the output voltage change of the sensor will be _____ volts.**

89. **Which of the following factors affect the accuracy of a pressure transducer?**

- A) Temperature of the environment
- B) Vibration of the surrounding area
- C) Humidity levels
- D) Altitude at which measurement is taken

90. **Select all true statements regarding the use of thermocouples in temperature measurement.**

- A) Thermocouples can measure temperature in a wide range.
- B) They require a reference temperature for accurate measurement.
- C) Thermocouples are unaffected by electromagnetic interference.
- D) They provide digital output directly.

91. **The resolution of a digital voltmeter used in an electrical circuit testing setup is _____ volts if it can distinguish between two measurements as close as 0.005 volts.**

92. **When calibrating an instrument, which parameters should be checked to ensure it meets the required standards? (Choose all that apply)**

- A) Linearity
- B) Hysteresis
- C) Repeatability
- D) Response time

93. **Calculate the maximum load a beam can withstand without yielding if the beam's yield strength is 250 MPa and the cross-sectional area is 0.01 m².**

94. **Which materials are most suitable for applications requiring high thermal conductivity?**

- A) Copper
- B) Glass
- C) Aluminum
- D) Rubber

95. **Select all the factors that influence the fatigue life of a mechanical component.**

   - A) Material grain size
   - B) Surface finish
   - C) Operating temperature
   - D) Color of the component

96. **The factor of safety for a structure designed to withstand a stress of 100 MPa, given that the ultimate tensile strength of the material used is 500 MPa, is _____.**

97. **Which of the following are considered when designing a gear system?**

   - A) Gear ratio
   - B) Load distribution
   - C) Material hardness
   - D) Color of the gears

98. **The bending moment at a section of a beam is calculated to be 5000 N·m. If the section modulus is 250 cm³, the bending stress at that section is _____ MPa.**

99. **Identify the appropriate heat treatment processes that can increase the toughness of steel.**

   - A) Quenching
   - B) Tempering
   - C) Annealing
   - D) Carburizing

100. **Select all true statements regarding the selection of a bearing for a high-speed application.**

   - A) Ball bearings are preferable.
   - B) Lubrication type is critical.
   - C) Load capacity must be prioritized over speed.
   - D) Roller bearings are unsuitable.

101. **The modulus of elasticity needed for a material that deforms 0.002 meters under a load of 10 kN and has a cross-sectional area of 0.05 m² is _____ GPa.**

102. **What are the typical considerations for choosing a composite material for automotive body parts?**

   - A) Weight reduction
   - B) Cost-effectiveness
   - C) Corrosion resistance
   - D) Aesthetic appeal

103. If the thermal expansion coefficient of a material is 12 x 10⁻⁶ /°C, the length change of a 2-meter long bar with a temperature increase of 50°C is _____ mm.

104. Which testing methods are appropriate for determining the mechanical properties of a new alloy? (Choose all that apply)

- A) Tensile testing
- B) Hardness testing
- C) Impact testing
- D) Electrical conductivity testing

105. Calculate the probability of drawing two aces in two consecutive draws from a standard deck of 52 cards without replacement.

106. Which of the following are properties of the normal distribution?

- A) Symmetry about the mean
- B) Mean equals median equals mode
- C) Skewness of zero
- D) All of these are properties

107. The variance of a dataset is 16. What is the standard deviation of this dataset?

108. Select all that apply: When performing a hypothesis test, which factors must be considered to determine the significance level?

- A) The confidence interval
- B) The risk of Type I error
- C) The power of the test
- D) The sample size

109. If the interquartile range of a set of data is 15, what is the range between the first quartile and the third quartile?

110. Determine the probability that a component manufactured in a plant passes a two-stage quality test, where the probability of passing the first stage is 0.8 and, given it passes the first stage, the probability of passing the second stage is 0.9.

Solutions

1 EXAM PREP

1. A) Center: (3, 2), Radius: $\sqrt{5}$

2. B) Minimum: 15, C) Maximum: 35

3. Click on the points (1,3).

4. 10.67

5. A) $\begin{bmatrix} 7 & -2 \\ -3 & 1 \end{bmatrix}$

6. 17

7. A) 7.07

8. B) 0.18
9. A) The mean is 5., C) The mode is 4., D) The standard deviation is 2.
10. (49.02, 50.98)
11. C) $\frac{1}{6}$
12. A, B, C, D

13. C) Public safety

14. B) Report the issue to the appropriate authorities
15. A, B, C
16. Education and experience

17. B) Review the project thoroughly before signing

18. B) 5 years

19. A, B, C

20. 30,696

21. B) 50%

22. Costs

23. B) 1.33

24. A, B, C

25. 4 μF

26. C) 4.8 ohms

27. A, B, D
28. Magnetic flux
29. C) 1000 N
30. A, C, D
31. $I = \frac{1}{3} bh^3$
32. D) 400 N
33. 554 N
34. 250 N
35. 600 Nm
36. C) Needs more data for assessment
37. Member forces depend on the truss configuration but can be found using methods such as the method of joints or sections. (General answer: Calculated by resolving the forces)
38. C) 2000 N
39. Centroid at (133.33 mm, 100 mm) from the bottom left corner of the L-shape
40. B) 4 m/s²
41. A, B, C
42. 0.80 Hz
43. D) 20 m
44. Gravitational acceleration
45. 29.43 m/s
46. A, C, D
47. C) 10 m/s²
48. Angular acceleration.
49. 150 J
50. A, B, C
51. 25 Hz
52. D) 200,000 Pa
53. A, B, C
54. 1×10^{-6}
55. 1.90476 ×10^−6 m
56. B) 17,657 kN
57. $20 \times 10^{10} \, Pa$
58. 500 Pa
59. 0.0034 m

60. Nc= 0.891×60= 53.46 RPM
61. 20MPa

$$\text{Shear Stress} = \frac{\text{Shear Force} \times Q}{I \times t}$$
$$\text{Shear Stress} = \frac{5000\,\text{N} \times 1000 \times 10^{-9}\,\text{m}^3}{200 \times 10^{-12}\,\text{m}^4 \times 0.05\,\text{m}}$$

62.

63. A) Copper
64. A, B, D
65. Dislocations.
66. B) Greater strength-to-weight ratio
67. A, B, D
68. Annealing
69. B) Hardness
70. A, B, D
71. C) 20400N
72. A, B, C, D
73. Weight

74. $\Delta P/m = \frac{f \rho v^2}{2D}$

75. B) 14 m/s
76. 150,000

77. $y_c = \left(\frac{Q^2}{g \cdot B^2}\right)^{1/3}$ 9.81m/s^2

78. $L = \frac{1}{2} \cdot \rho \cdot v^2 \cdot C_L \cdot A$

79. Square of the velocity

80. $Q = v \cdot A = v \cdot \left(\frac{\pi D^2}{4}\right)$

81. $F_D = \frac{1}{2} \cdot \rho \cdot v^2 \cdot C_D \cdot A$

82. A) 49 kPa

83. B) First Law of Thermodynamics

84. A, B

85. Carnot

86. C) 89.3 kJ

87. Internal energy change (ΔU) when a system absorbs heat Q and does work W is:
$\Delta U = Q - W$
$\Delta U = 500\,J - 300\,J = 200\,J$

88. A, B,

89. High-temperature and low-temperature

90. A) 400 J/kg·°C

91. 501.6kJ

92. A, B, D

93. Second

94. A) 9.1

95. C) Radiation

96. A, B, D

97. Fourier's law

98. C) 48 W

99. A, C, D

100. 1

101. $Q = 0.9 \times 5.67 \times 10^{-8} \times 2 \times 500^4$

102. 45 W

103. B) Ammeter

104. A, B, C

105. Thermocouple

106. B) 900 ohms

107. A, B

108. C) 0.316 m/s

109. A, B, C

2 EXAM PREP

$$\frac{x^2}{4} + \frac{y^2}{16} = 1$$

1. A)
2. A, B, D
3. 33.51
4.
$$z = -1.86 - 3.37i$$
$$z = -1.14 - 0.63i$$

5. Expected value is 2.1, and the variance is 0.49.
6. A) 4,2
7. A) $\frac{1}{6}$
8. A, C, D
9. Binomial distribution
10. The mean is 5.0, and the variance is 4.0
11. D) All of the above
12. B) The public
13. B, D
14. Education and experience
15. D) Refuse the gift and report the offer to their professional association
16. A, B, D
17. B) 5 years
18. A, B, D
19. Discount
20. A) $6,691.13
21. A, B, C, D
22. C) 0.4 A
23. A, B, C
24. Faraday's law
25. C) 45 W
26. A, B, C, D
27. A) A=900 N, B=900 N
28. A, B
29. Weighted average
30.
$$I_{\text{rectangle}} = \tfrac{1}{3} \times \text{base} \times \text{height}^3 = \tfrac{1}{3} \times 200 \times 400^3$$
$$I_{\text{semicircle}} = \tfrac{1}{8}\pi r^4 + A \times d^2$$

31. B) 0.6

32. A, C, D
33. B) 2500 Nm
34. A column's stability under a compressive load depends on its **slenderness ratio** and **material properties**.

35. $M_{max} = \dfrac{P \times L}{8}$
36. C) 30 kPa
37. C) 30 m/s
38. A) The tension in the vertical cable equals the weight of the mass.
39. Approximately 0.47.
40. B) 2.21 Hz
41. D) 5000 Ns
42. A, B, C, D
43. 2.46 s
44. $F = \dfrac{mv^2}{r}$, 50 N
45. $R = \dfrac{v^2 \sin(2\theta)}{g}$
46. 3.56 Hz
47. B) 50 MPa
48. A, B, D
49. $\sigma = \dfrac{My}{I}$
50. A) 0.12 J
51. $\tau = \dfrac{F}{A}$, 70.74 MPa
52. A, C, D
53. Euler's
54. $\sigma = E\alpha\Delta T$, 126 MPa
55. $\theta = \dfrac{TL}{JG}$, 0.018 rad
56. A) 3.75 MPa
57. B) Copper
58. A, B, C, D
59. Rockwell
60. C) 2.5 g/cm³

61. A, B, C
62. Carburizing
63. C) Tungsten
64. A, B, D
65. B) 0.013
66. A, B, C
67. Navier-Stokes
68. B) 21.02 kW
69. 19.2 mm
70. A, B, C
71. Joukowsky
72. B) , 0.204 m/s
73. 882 N
74. C) Transition
75. A) 0 J
76. B, D
77. Carnot
78. B) 50%
79. 200 J/(kg°C)
80. A
81. Second
82. C) 500 J
83. 50°C
84. C) 700 J
85. A) 8 W/m²
86. A, B
87. Stefan-Boltzmann

$$Q = \epsilon \sigma A T^4 \Delta t$$

88.
89. A) 453.24 W
90. A, C, D
91. Nusselt
92. A) 1428.6 J
93. C) 40°C
94. A, C
95. Tolerance
96. B) 16 dB
97. A, B, D
98. Faraday's
99. C) 675 Nm
100. A, B, C
101. Safety
102. A) 0.533 m
103. A, B, C, D

104. Von Mises
105. 127.32 Nm
106. A, C, D
107. Strength
108. B) 200 MPa
109. A, B, C, D
110. Finite Element

3 EXAM PREP

2.
   1. $y(0) = C_2 + 3 = 0 \Rightarrow C_2 = -3$
   2. $y(1) = C_1 \sin(1) - 3\cos(1) + 3 = 0$
3. A) $2\pi i$
4. After two iterations, the solution is approximately (0.3, 1.47, 2.676)
5. B) 2520
6.
$$\left.\frac{\partial f}{\partial x}\right|_{(1,2)} = 2y + 2xe^{xy} = 2(2) + 2(1)e^2 = 4 + 2e^2$$

$$\left.\frac{\partial f}{\partial y}\right|_{(1,2)} = x^2 + xe^{xy} + 3y^2 = 1^2 + 1e^2 + 3(2)^2 = 1 + e^2 + 12 = 13 + e^2$$

7. E(X)=0 Var(X)=2
8. D) 3/8
9. A, B, C
10. Central Limit
11. 5
12. B) 1/15
13. B) Report the issue to the appropriate authorities and stakeholders.
14. A, B
15. Safety
16. B) Acknowledge the mistake and take responsibility.
17. A, B, D
18. B) $12,762.82
19. A, B, D
20. Internal Rate of Return (IRR)
21. C) $19,964
22. A, B, C
23. A) $3.36 \times 10^{-5}\ \Omega$
24. A, B, C
25. Ampère's

26. A) 48 V

27. A, B, C

28. D) 250 N

29. The centroid is approximately 225 mm from the bottom of the T-shaped object.

30. C) $0.356 \times 10^6 \text{ mm}^4$

31. $F_{\max} = \mu_s \times m \times g \times \cos(\theta)$

The maximum load without slipping is approximately 0.59 times the normal force.

32. For a 600 kg mass suspended by two cables at angles of 30° and 60° to the vertical, the tension in each cable is approximately **4308.85 N**. This assumes that the load is symmetrically distributed between the two cables.

33. C) 266.67 N

34. The bending moment at the midpoint is 1250Nm

35. $P_{\text{critical}} = \dfrac{\pi^2 \times E \times I}{(2L)^2}$

36. In the simplified model of a symmetric square truss with a central load of 1200 N, each member will experience a force of approximately 300 N

37. A) 2450 N

38. B) 38.7 m/s

39. Assuming symmetric loading, each cable supports part of the vertical load of 150 kg×9.81 m/s^2

40. C) 0.577

41. $\omega_d = 13.86 \text{ rad/s}$

42. A) -30,000 kg·m/s

43. 48,000 J

44. D) 3.28 s

45. D) 450 N

46. 35.36 m/s

47. A=0.03 m

48. B) 60 MPa

49. C) 0.0667 mm

216

50. C) 131.17 MPa
51. B) 1000 N
52. D) 4.25 x 10^6 mm^4
53. B) 5654.7 rad/s
54. 0.25
55. A) 72.36 MPa.
56. B) 250 J
57. A) 12.73 MPa
58. 0.000476
59. 0.167 mm.
60. C) 1527.89 MPa
61. A) 2.0 MPa.
62. B) 3.68 mm
63. A) 18.75 MPa
64. 4306.43 kN
65. D) 450 MPa.
66. C) 6264.18 N
67. 326 Pa/m
68. B) 0.00136 m³/s
69. C) 9.81m/s^2
70. 326.45 meters
71. C) 0.7
72. A), B)
73. 46.14 Newtons.
74. B) 0.0023 m/s
75. A) The gravitational constant, B) The density of water, D) The mechanical losses in the pump
76. C) 8.2 m/s
77. 2 J/K
78. B) 25%
79. A) The internal energy change over one cycle is zero, C) The system returns to its original state at the end of the cycle, D) Energy is neither created nor destroyed.
80. 1254 kJ
81. A) 0.83
82. A) Determining the maximum efficiency of heat engines, C) Predicting the direction of heat transfer, D) Evaluating the performance of refrigeration cycles.
83. 2175 J/mol·K
84. A) -50 J
85. A) Maximizing the boiler pressure, B) Minimizing the condenser pressure, C) Increasing the superheat temperature.
86. 500 J
87. A) Sensors, B) Actuators, C) Set point
88. 0.25 volts

89. A) Temperature of the environment, B) Vibration of the surrounding area, C) Humidity levels, D) Altitude at which measurement is taken
90. A) Thermocouples can measure temperature in a wide range, B) They require a reference temperature for accurate measurement
91. 0.005 volts
92. A) Linearity, B) Hysteresis, C) Repeatability
93. 2.5 kN
94. A) Copper, C) Aluminum
95. A) Material grain size, B) Surface finish, C) Operating temperature
96. 5
97. A) Gear ratio, B) Load distribution, C) Material hardness
98. 20000 MPa
99. B) Tempering, C) Annealing
100. A) Ball bearings are preferable, B) Lubrication type is critical
101. 200 GPa
102. A) Weight reduction, B) Cost-effectiveness, C) Corrosion resistance
103. 1.2 mm
104. A) Tensile testing, B) Hardness testing, C) Impact testing
105. 0.0045
106. D) All of these are properties
107. 4
108. B) The risk of Type I error, C) The power of the test, D) The sample size
109. 15
110. 0.72

Mastering Test Strategies: Time Management, Identifying Correct Solutions, and Avoiding Pitfalls

Grasping the technical material is only a portion of what's required for success in the exam you're preparing for. It's equally critical to refine your exam-taking strategies. Excelling in them will empower you to handle the exam effectively. In this guide, I will outline the key techniques that have helped hundreds of students and candidates to achieve certification. These include adept time management, precise identification of correct answers, and the elimination of common mistakes that could hinder your progress.

Time Management Techniques

Time is an invaluable asset during the FE exam, and managing it wisely can mean the difference between passing and failing. To illustrate the critical role of time management, let's consider two hypothetical students:

Student A knows every topic thoroughly but lacks time management skills. During the exam, this student spends excessive time on complex problems, feeling compelled to utilize their extensive knowledge. As a result, they run out of time and miss several easier questions that could have been quickly solved.

Student B has a good but not exhaustive grasp of the material. However, Student B has practiced efficient time allocation. Understanding that once they move past a question they cannot return, they decide quickly whether a question is within their immediate ability to answer. If it appears too complex or time-consuming, they make an educated guess and move on, ensuring they have enough time to address all questions, prioritizing those they can definitely solve.

In this scenario, even though Student A may have deeper knowledge, Student B likely achieves a higher score due to their strategic approach.

Here's how you can optimize your time:

1. **Understand the Exam Format**: Before the exam, review all provided materials to understand the distribution of topics and the format of questions. This knowledge allows you to strategize your approach based on the section or subject areas, adjusting your pace to match the complexity of each section.

2. **Prioritize Questions**: Not all questions are created equal. Begin with questions that you find easier or are more confident about would be nice. Unfortunately this is not allowed when it comes to FE Exam. In this context questions must be answered in sequence without skipping, here are some strategies to tackle difficult questions effectively:

    - **Analyze the Answers**: Sometimes, just reading through the answers provided can give clues about what the question is really asking.
    - **Choose the Most Probable Answer**: If you're unsure, select the answer that seems most likely based on your knowledge and understanding. It's often useful to trust your first instinct unless you find clear evidence to change your mind as you analyze the question further.
    - **Decide and Move On**: Once you have made a choice, it's important to move on to the next question without dwelling on the possibility of having made a mistake. Second-

guessing yourself can waste precious time and diminish your focus on subsequent questions.

3. **Set Time Limits:** Allocate specific times to each question or section. If a question takes too long, move on repeating the concepts just seen. For instance, if you have 6 hours for 110 questions, plan to spend no more than about 3 minutes per question. Use a watch to keep track of time discretely during the exam and adjust your pace as needed to ensure you cover all questions.
4. **Use of Breaks:** Schedule brief mental breaks to refresh your focus, especially during longer testing sessions. This can involve simple stretching, deep breathing, or just closing your eyes for a minute. Planned breaks prevent fatigue and maintain mental clarity, contributing to better decision-making.
5. **Practice with Simulated Exams:** Regularly take full-length practice exams under timed conditions to build your stamina and get accustomed to managing the allocated time effectively. This practice will help you understand how quickly you need to move through the exam to answer all questions without rushing.

   Recreate the environment in which you'll take the test. This involves mimicking every detail of the exam day—dress as you would on that day, follow your intended morning routine, and manage your diet and sleep just as you plan to for the actual exam. Treating a practice session as the real exam day is a powerful strategy to condition your mental and emotional responses, helping you gauge your readiness under true test conditions.

This process also allows you to study your resistances and the behaviors you adopt. Do you experience anxiety? When does this emotion develop? Right upon waking, or when you step into the classroom for a simulation?

Does this lead to a lack of focus and panic, or can you manage the anxiety effectively? Answering these questions can help you understand where you need to focus your efforts more. Take, for example, the fact that you don't wake up anxious but start to sweat as you approach the exam.

I have good news for you—it's completely normal and can be resolved. The discomfort you feel is simply because you are not used to the situation. In this case, what you need to do is continue these simulations, and you will see that the discomfort you experience will gradually decrease each time.

Conversely, consider a person who might wake up feeling anxious. Here's another little secret you already know but probably ignore: it's perfectly normal and happens to everyone. Here, insecurity is speaking for you; what I have seen greatly improve this situation is an even deeper understanding of the subject, so reviewing and completely eliminating any gaps will allow you to wake up relaxed and ready to face the situation.

Now, let's consider a student who wakes up anxious, enters the classroom, and feels even more discomfort, but for some "strange" reason, manages to maintain focus and answer the questions well, achieving excellent results. You might wonder, but I assure you, these students are concerned about the situation just described. For all those who find themselves in this scenario, it is about becoming aware that you will experience these emotions and still be able to pass the exam.

A crucial part is understanding where you stand. Performing these simulations allows you to answer questions and, based on where you categorize yourself, you are able to work on your situation.

Here's 10 questions to answer to find your position and start improving:

1. What specific situations trigger my anxiety the most during exams or simulations?
    - Identify their stress triggers clearly.

2. How do I usually respond physically and mentally when I start feeling anxious?
    - Recognize developing coping mechanisms.
3. What relaxation techniques have I tried, and which ones have been effective for me?
    - Evaluate and choose suitable relaxation strategies.
4. How can I incorporate regular anxiety management exercises into my daily routine?
    - Integrate stress management into your daily life.
5. Do I have a study plan that allows for breaks and time to decompress, and how can I improve it?
    - Improve your study schedules to reduce stress.
6. What are my most common doubts or fears about the exam, and what evidence do I have that counters these fears?
    - Rationalize fears and boost confidence by confronting irrational beliefs.
7. Who in my support network can I talk to when I'm feeling overwhelmed?
    - Remember to utilize all the help you need effectively.
8. What positive affirmations can I use to boost my confidence right before and during an exam?
    - Use positive psychology to improve performance.
9. How can I simulate exam conditions more effectively to become accustomed to the pressure?
    - Strategic practice under realistic conditions to reduce anxiety during the actual exam.
10. What are my strategies for dealing with questions that I find difficult or confusing during the exam?
    - Plan and prepare for handling challenging questions without panic.

Identifying Correct Solutions

As you delve into the core of the exam, identifying the correct solutions becomes crucial. The ticking of the clock and the plethora of choices can make this task seem daunting, even though you can apply everything we've previously discussed. With the right strategies and a calm, calculated approach, you can turn this challenge into a manageable and even rewarding part of your experience. In this section, I will guide you through a series of proven strategies that not only enhance your accuracy but also strengthen your confidence in selecting the right answers. These techniques are designed to help you methodically dissect each question. Whether you are faced with a straightforward question or a complex problem, the following strategies will equip you with the necessary tools to tackle each scenario.

Here are strategies to improve your accuracy:

1. **Look for Keywords**: Pay attention to keywords in both the questions and answers. Terms like "always," "never," "only," etc., can be crucial in understanding what the question specifically requires. They often define the scope and are directly linked to the correctness of an answer. Keywords help in focusing on what the question explicitly asks and avoid traps set by commonly used distractors.
2. **Use Logical Deduction**: When faced with tough questions where the answer isn't immediately apparent, use logical reasoning to deduce the most likely correct answer. This involves linking

concepts from different parts of the syllabus that relate to the question, providing a well-rounded approach to solving complex problems.

3. **Review Educated Guesses**: If you must guess (after narrowing down the options), make an educated guess rather than a random selection. This means choosing an answer based on partial knowledge and logical assumptions rather than on whim.

4. **Prioritize Time Management**: Ensure you adhere to the concepts we've previously discussed. Allocate specific amounts of time to each question based on its difficulty or the points it offers. This strategy helps you avoid spending too much time on the more difficult questions at the expense of the easier ones.

5. **Utilize Diagrams and Visuals:** Whenever possible, quickly draw diagrams or sketches to visualize complex problems. This can clarify the question and lead to a quicker resolution of the problem. This method is excellent for clarifying when you are confused. Our minds reason in images, and if you want to validate this thoroughly valid thesis, think about what you are reading right now. What happens is that if I say "house," an image of a house appears in your mind—it could be a drawing, an image, a photo, or even your room. Everyone associates each concept with the image they find most aligned with it. Having an image, a diagram helps to clearly understand what is being discussed, and in the case of a question, this is crucial to ensure that the correct answer is given.

6. **Break Down Complex Questions**: Break complex questions into smaller, more manageable parts. Analyzing each part separately can simplify the problem and lead to a clearer understanding of the question. Have you ever felt overwhelmed by having too much to do? I believe you have. The key lies in thinking about one step at a time. This strategy allows you to simplify the question and respond one piece at a time. Here, complexity is eliminated, allowing you to make progress.

7. **Use Backward Elimination**: For multiple-choice questions, work backward from the answer choices to the question. This can help validate the correctness of an answer by aligning it with the question's requirements.

   Imagine a question asking, "Which of the following options is a non-renewable resource?" with the choices being A) Coal, B) Solar Energy, C) Wind, D) Hydroelectric Power.

   Given that renewable resources are those that naturally replenish, Coal, as option A, is a fossil fuel and not naturally replenishing, making it a non-renewable resource. Remembering that Solar Energy (option B) harnesses the sun, which is perpetually renewable, invalidates this choice. Similarly, both Wind (option C) and Hydroelectric Power (option D) are powered by naturally recurring phenomena.

   It's apparent that option D, Hydroelectric Power, involves water cycles which are naturally renewable, ruling it out as well. Thus, by process of elimination, you find that Coal, option A, is the correct answer because it doesn't align with the defining characteristics of renewable resources. This backward elimination not only confirms the correct answer but also reinforces your understanding of the concept by excluding incompatible options.

8. **Refer to real-world applications**: Try to connect theoretical questions to scenarios or applications from the real world. Along with using visual aids, linking topics to actual facts is an extremely powerful method, primarily for memorization and learning, but especially for testing answers. This context can often provide clarity and a deeper understanding of the correct approach.

9. **Seek consistency in responses**: When multiple questions pertain to a single scenario or dataset, ensure that your answers are consistent across them. Inconsistencies can sometimes signal misunderstandings or calculation errors.

    Here's a compelling example in the context of mechanical engineering:

    Scenario: You are provided with data on the thermal expansion of a steel beam in response to varying temperatures during a design simulation.

    Question: If the initial length of a steel beam is 10 meters at 20°C, and the coefficient of thermal expansion is $1.2 \times 10^{-5} \,°C^{-1}$, calculate the length of the beam at 50°C.

    Question follow-up: Considering your previous answer, calculate the stress induced in the beam if it is constrained and cannot expand, given that the modulus of elasticity for steel is 200 GPa.

    In this scenario, if you calculate an expanded length of the beam at 50°C for the first question, you must use this length to determine the stress in the second question. Any discrepancy in using different lengths for these calculations will introduce inconsistency, suggesting a misunderstanding of the concept or an error in your calculations. Therefore, ensure your answers logically follow from one another, reflecting a coherent understanding of the scenario's physics.

10. **Cross-Verify with Other Questions**: Occasionally, information or data provided in one question can help answer another. Keep an eye out for such opportunities as they can provide easy marks.
11. **Maintain a Calm Composure**: Keep your stress levels in check. Anxiety can cloud judgment and impede your ability to process information efficiently. Taking deep breaths or pausing briefly as we discussed previously can help maintain focus.
12. **Reflect on Practice Exams**: Use insights gained from practice exams to identify patterns in the types of errors you typically make. Addressing these patterns can reduce repeated mistakes and improve overall performance during the actual exam.
13. **Learn from Your Mistakes**: Every mistake you make is a question you won't get wrong again, if you take notes on why you answered incorrectly and what the correct answer is. Humans learn more from their failures than their successes. Apply this concept to all the simulations you do, and you will see a) the areas where you have more difficulty, b) you will master the concepts you previously got wrong.

Avoiding Common Pitfalls

Be aware of the traps that can derail your exam performance:

1. **Over-Preparation**: While it's important to be thoroughly prepared, overdoing it can lead to burnout. Ensure you balance study with breaks to keep your mind fresh. Find a balance in your study routine.
2. **Misreading Questions**: Under exam pressure, it's easy to misread questions or miss out on crucial details. Always take a moment to read each question carefully and consider reading them more than once, highlight key terms and conditions to ensure you understand exactly what is being asked .
3. **Time Mismanagement**: Poor time management is one of the most common pitfalls in exam settings. It can result in rushing through questions or, conversely, spending too much time on

challenging ones, leaving insufficient time to complete the exam. To avoid this, use practice exams to develop a sense of pacing. Allocate a specific amount of time for each question or section and stick to it during the exam. I'll repeat, regularly check the clock and adjust your pace accordingly to ensure that you have enough time to address all questions.

4. **Ignoring Simple Questions**: Sometimes, in an attempt to tackle more complex problems, candidates overlook simpler questions or commit careless errors on them. Ensure that every question is treated with care, checking even the simplest answers for accuracy.

5. **Underestimating the Importance of Practice Exams**: Practice exams are not just for reinforcing content knowledge; they also help in acclimating to the pressure and timing of the real exam. Regularly taking timed practice exams can reduce anxiety and increase your comfort level with the exam format.

6. **Neglecting Mental and Physical Health**: The rigors of exam preparation can sometimes lead to neglect of physical and mental health. Maintaining a healthy diet, getting regular exercise, and ensuring adequate sleep are all crucial for peak cognitive function. Remember, a healthy body contributes to a sharp mind.

Practical Application of Test Strategies

To integrate these strategies into your exam preparation, consider simulating the exam environment:

Practice Under Timed Conditions: As I suggested, consistently practicing under timed conditions is crucial. Set up a timer to mimic the actual exam duration. This not only accustoms you to the pressure of the ticking clock but also sharpens your ability to allocate time efficiently across different sections of the exam.

Review and Adjust: After completing each practice test, take the time to thoroughly review your answers. This review should focus not just on the questions you got wrong, but also on analyzing the time spent on each question. Identify patterns or specific types of questions where you consistently lose time or make errors. Are there certain topics or question formats that consistently challenge you? Use this insight to adjust your study plan—focus more on these weaker areas.

Create a Feedback Loop: Implement a systematic feedback loop into your study routine. This involves recording your performance on each practice test, including the types of errors made and the topics that need more attention. Over time, this will help you track your progress and pinpoint areas that require more focused review or a different approach.

Peer Reviews and Group Studies: Occasionally, involve a peer or a study group in your practice sessions. After taking a timed test, exchange papers and critique each other's work. This can provide new perspectives on problem-solving and help uncover mistakes that you might not have noticed on your own

Use of Technology and Resources: Utilize available technology and resources effectively. There are various online platforms and apps designed to simulate exam conditions and provide realistic practice questions. These tools often offer immediate feedback and detailed explanations, which can be invaluable for understanding complex topics and improving problem-solving skills.

Remember, effective preparation is not just about covering all topics but also about preparing mentally and strategically to handle the exam format and pressures. By practicing these techniques, you are setting yourself up for success, ensuring that you not only perform optimally on the exam day but also turn these strategies into habits that will benefit your future professional engineering career.

Chapter 17: Beyond the FE Exam: Pathways to Professional Engineering

Steps After the Exam: Licensure and Lifelong Learning

Passing the FE Mechanical Exam is just the beginning of a thrilling journey into the world of professional engineering. The path ahead is rich with opportunities for growth, innovation, and leadership. Here's what you need to focus on next: achieving your licensure and committing to lifelong learning.

Gaining Professional Licensure

Once you've passed the FE Exam, the next step is to become a licensed Professional Engineer (PE). This credential is your gateway to higher responsibility, increased autonomy in your work, and often, a better compensation package.

1. **Work Experience**: Typically, you will need a few years of engineering work experience under the supervision of a licensed PE. This period allows you to apply your academic knowledge in real-world settings and refine your technical and professional skills.

2. **PE Exam**: After meeting the experience requirement, you'll be eligible to sit for the Principles and Practice of Engineering (PE) Exam in your specific discipline. Passing this exam is crucial as it validates your expertise and commitment to the highest standards of engineering.

3. **Licensure Application**: With the PE exam cleared, you can apply for licensure through your state's engineering board. Each state may have different requirements, so it's vital to ensure you meet all the criteria to be granted your license.

Embracing Lifelong Learning

One thing that is certain in this field is that it will never stop is 'innovation and progress. Because this statement is true, it is equally true that you will continue to learn throughout your life. This means continuous learning and professional development.

Here's how you can keep growing professionally and personally:

1. **Advanced Degrees and Certifications**: Consider pursuing a master's or doctoral degree in a specialization of interest. Certifications in new technologies, project management, or leadership can also enhance your credentials and marketability.

2. **Professional Development Courses**: Many engineering societies offer courses that help you stay updated with the latest technologies and best practices. These courses are not only educational but also a great way to earn Professional Development Hours (PDHs) required for maintaining your PE license.

3. **Attending Conferences and Workshops**: These gatherings are goldmines for learning and networking. They provide insights into cutting-edge research, emerging industry trends, and practical skills that you can apply in your day-to-day work.

4. **Reading and Research**: Stay curious and informed. Regularly read engineering journals, books, and publications. Engage with the latest research in your field to spark innovation in your projects.
5. **Networking**: To open doors to new opportunities and collaborations, building a strong professional network may be the best solution. Join engineering associations, attend industry meetups, and connect with peers and mentors who can offer guidance and support.

Remember that the trick is in applying, not just reading this information. These tips are accessible to anyone, but you have to put them into practice to succeed in scale-up. The only way to do that is to take action!

Building a Career: Opportunities, Networking, and Continuing Education

Opportunities in Mechanical Engineering

The field of mechanical engineering offers diverse opportunities across various industries including aerospace, automotive, manufacturing, and energy.

Here's few samples:

- **Aerospace Engineer**: Focuses on the design and production of aircraft, spacecraft, satellites, and missiles.
- **Automotive Engineer**: Involves designing, manufacturing, and testing vehicles including cars, trucks, motorcycles, and heavy equipment.
- **Biomedical Engineer**: Combines engineering principles with medical and biological sciences to design and create equipment, devices, computer systems, and software used in healthcare. This includes innovations like artificial organs, prostheses, and medical instrumentation.
- **Construction Engineer**: Engages in the design, planning, construction, and management of infrastructure including roads, tunnels, bridges, airports, railways, facilities, buildings, as well as water supply and sewage treatment systems.
- **HVAC Engineer**: Specializes in heating, ventilation, and air conditioning systems, focusing on the design and implementation of these systems to improve energy efficiency and comfort in buildings.
- **Manufacturing Engineer**: Designs and optimizes manufacturing processes for efficiency and safety.
- **Mechanical Engineer**: Often found in industries like manufacturing, aerospace, automotive, chemical, and construction, focusing on designing and developing mechanical and thermodynamic equipment.
- **Robotics Engineer**: Works with robotics systems that apply concepts
- **Energy Engineer**: Focuses on the production and supply of energy through natural resources, such as the extraction of oil and gas, as well as from renewable or sustainable sources of energy, including biofuels, hydro, wind, and solar power.

There are a lot of interesting fields to grow in, don't you think? Many fields are in development. Many changes potentially represent the future. Just think about how technologies are evolving today: very fast. The medical field is an ever-evolving field, and it also combines very well with the field of robotics. Consider fully immersing yourself in one of these areas and making a career out of it. In a few years, you could find yourself as a leader at global companies.

Should you be interested in this or otherwise want to grow within a community led by engineers who face progress head-on and achieve it, here are some tips that might come in handy:

1. **Specialize**: Consider specializing in high-dem and areas such as robotics, renewable energy, or computational fluid dynamics. Specialization makes you more attractive to employers looking for experts in specific fields.
2. **Innovation and Entrepreneurship**: Leverage your expertise to innovate. Whether developing new products or improving existing ones, innovation drives career growth and satisfaction. Additionally, entrepreneurial ventures can open new pathways and transform your career.
3. **Global Opportunities:** Mechanical engineering skills are globally relevant. Look for opportunities abroad to work on international projects or with multinational companies. This exposure can significantly enhance your professional development and personal growth.

Networking for Career Advancement

Networking is a powerful tool for career development. It connects you with industry professionals, exposes you to job opportunities, and provides access to mentors who can guide your career path. And I consider it so important that I should devote a separate section to it.

Here is how you can increase vosltro networking:

1. **Professional Associations**: Join organizations like ASME (American Society of Mechanical Engineers) or SAE International. These associations offer networking events, professional development resources, and access to the latest industry trends.
2. **Conferences and Seminars:** Regularly attend industry conferences and seminars. These events are not only educational but also provide a platform to meet industry leaders and peers.
3. **Online Platforms:** Leverage professional networking platforms such as LinkedIn to engage with industry experts, participate in pertinent groups, showcase your accomplishments, and gain insights from others.
4. **Networking Opportunities:** Get out of your comfort zone and push yourself to find people who share your calling and align with your goals. You can literally start with your university or your field of study.
5. **Events and Gatherings:** Do a search on social media, and you will find it is full of communities that share dreams like yours. Join them and expand your knowledge.
6. **Digital Communities:** Organize events to build a community. Start with the same social groups and ask to organize an outing where you can compare ideas and grow together

Once again. Take action! Take one of these tips and apply it right away. One is enough for you to get to your first contacts. Once done move on to the next

Continuing Education

Continual learning is a must in keeping up with technological advances and remaining competitive in the field. As we have already mentioned, embracing lifelong learning is what differents enginiers for professional and personal growth. Here are some specific ways to achieve this:

1. **Advanced Degrees and Certifications:**
    - **Advanced Degrees:** Pursuing an advanced degree, such as a Master's or Ph.D., can open up advanced technical and management positions that might otherwise be inaccessible.
    - **Certifications:** Earning certifications in specific technologies, project management, or lean manufacturing can significantly enhance your skills and make you stand out in the job market. Certifications demonstrate your dedication to staying current with industry standards and best practices.
2. **Professional Development Courses:**
    - **Engineering Societies:** Many engineering societies offer courses that help you stay updated with the latest technologies and best practices. These courses are educational and a great way to earn Professional Development Hours (PDHs) required for maintaining your Professional Engineer (PE) license.
    - **Industry-Specific Training:** Engage in training programs tailored to your field. These programs can provide targeted skills and knowledge that are directly applicable to your current or desired job roles.
3. **Online Courses and Workshops:**
    - **Digital Learning Platforms:** Platforms like Coursera, EdX, and others offer a wide range of courses that can help you update your skills or learn new ones relevant to evolving industry needs. These courses are flexible and can be completed at your own pace, making them ideal for working professionals.
    - **Workshops:** Participating in workshops, whether in-person or virtual, can provide hands-on experience and direct interaction with experts in the field. Workshops often focus on practical skills and real-world applications, which can be immediately beneficial in your job.
4. **Attending Conferences and Seminars:**
    - **Learning and Networking:** Conferences and seminars are invaluable for learning about cutting-edge research, emerging industry trends, and practical skills. They also offer excellent networking opportunities, allowing you to connect with industry leaders, peers, and potential collaborators.
    - **Professional Growth:** These events can inspire new ideas and provide a broader perspective on your field, contributing to both your professional and personal growth.
5. **Reading and Research:**
    - **Staying Informed:** Regularly reading engineering journals, books, and publications keeps you informed about the latest advancements and research in your field. This habit fosters a culture of continuous learning and can spark innovation in your projects.
    - **Engagement:** Stay up-to-date with the latest research and advancements to remain at the forefront of your field. This proactive strategy can result in improved decision-making and more creative solutions in your work.

Building a Career Post-Exam

The transition from passing the FE exam to building a successful career requires a strategic and multifaceted approach. As with the exam, it is necessary to have a strategy that leads to achievement and to plan. Here is a detailed guide to help you navigate this journey effectively:

1. Mentorship

Find a Mentor

- **Importance of Mentorship:** The mentor is a most important figure. It represents a role model from whom to learn, always ready to give advice and correct you. He can provide invaluable insights, advice, and support. For instance, you have the opportunity not to repeat the mistakes he has made before thus facilitating and accelerating your career development. They help navigate the complexities of the industry, offer critical feedback on your work, and open doors to opportunities.

How to Find a Mentor

- **Networking Events:** Attend industry conferences, seminars, and workshops where you can meet experienced professionals.
- **Professional Associations:** Join organizations like ASME (American Society of Mechanical Engineers) or other relevant bodies where mentorship programs may be available.
- **LinkedIn Connections:** Use LinkedIn to connect with industry leaders and professionals. Look for individuals whose careers you admire and reach out to them with a thoughtful message expressing your interest in learning from their experiences.

2. Career Planning

Set Clear Career Goals

- **Short-Term Goals (1-3 years):**
    - **Job Roles:** Identify entry-level positions that align with your interests and strengths.
    - **Skills Acquisition:** Focus on acquiring skills that are in high demand within your industry. This could include technical skills like CAD software proficiency or soft skills like project management.
    - **Certifications:** Obtain relevant certifications such as Six Sigma, PMP (Project Management Professional), or specific technical certifications.
- **Long-Term Goals (5-15 years):**
    - **Career Roadmap:** Develop a career roadmap outlining key milestones and the steps needed to achieve them. This could include roles like senior engineer, project manager, or technical director.
    - **Advanced Degrees:** Consider pursuing advanced degrees such as a Master's or Ph.D. to open up higher-level technical and management positions.
    - **Leadership Roles:** Aim for leadership or executive positions within your field, such as becoming a department head or a VP of Engineering.

Tools and Resources

- **Career Counseling:** Utilize career counseling services offered by your university or professional associations.
- **Workshops and Seminars:** Attend career planning workshops to gain insights on how to effectively plan your career trajectory.

3. Work-Life Balance

Manage Your Time Effectively

- **Prioritize Tasks:** Use tools like to-do lists, calendars, and project management software to prioritize tasks and manage your time efficiently.
- **Set Boundaries:** Establish clear boundaries between work and personal life. This might involve setting specific work hours and sticking to them.
- **Know When to Disconnect:** Ensure you take time to disconnect from work to recharge. This can help prevent burnout and maintain a healthy work-life balance.

We all have 24 hours and 7 days a week. Practicality lies in being able to organize yourself and find what works for you, not just following a copied and pasted template. This comment should not be confused with the uneasiness you feel when trying something new and incorporating it into your routine.

As we have studied, you need to test and see if it works or not. Give yourself time to determine if the routine you choose doesn't work for you or if you can get used to it. Remember that flexibility is a component to consider since we are not machines.

Activities to Recharge

- **Hobbies and Interests:** Engage in activities and hobbies outside of work that you enjoy. This could include sports, reading, or any other pastime that helps you relax.
- **Exercise and Health:** Maintain a regular exercise routine and focus on a healthy diet to keep your mind and body in top condition.
- **Family and Social Life:** Spend quality time with family and friends. Building strong personal relationships can provide emotional support and enhance your overall well-being.

4. Practical Experience

- **Internships and Co-ops:** Gain practical experience through internships and cooperative education programs. This hands-on experience is invaluable and can often lead to full-time job offers.
- **Project Involvement:** Volunteer for projects at work that allow you to apply your skills and gain new experiences. This can also demonstrate your initiative and dedication to your employer.

By following these strategies and taking a proactive approach, you can transition successfully from passing the FE Exam to building a rewarding and successful career in the field of mechanical engineering. Remember, the key is to continuously seek opportunities for growth, apply the knowledge you gain, and take action towards achieving your career goals.

Made in the USA
Middletown, DE
30 August 2024

59975509R00128